Springer Theses

Recognizing Outstanding Ph.D. Research

Aims and Scope

The series "Springer Theses" brings together a selection of the very best Ph.D. theses from around the world and across the physical sciences. Nominated and endorsed by two recognized specialists, each published volume has been selected for its scientific excellence and the high impact of its contents for the pertinent field of research. For greater accessibility to non-specialists, the published versions include an extended introduction, as well as a foreword by the student's supervisor explaining the special relevance of the work for the field. As a whole, the series will provide a valuable resource both for newcomers to the research fields described, and for other scientists seeking detailed background information on special questions. Finally, it provides an accredited documentation of the valuable contributions made by today's younger generation of scientists.

Theses are accepted into the series by invited nomination only and must fulfill all of the following criteria

- They must be written in good English.
- The topic should fall within the confines of Chemistry, Physics, Earth Sciences, Engineering and related interdisciplinary fields such as Materials, Nanoscience, Chemical Engineering, Complex Systems and Biophysics.
- The work reported in the thesis must represent a significant scientific advance.
- If the thesis includes previously published material, permission to reproduce this must be gained from the respective copyright holder.
- They must have been examined and passed during the 12 months prior to nomination.
- Each thesis should include a foreword by the supervisor outlining the significance of its content.
- The theses should have a clearly defined structure including an introduction accessible to scientists not expert in that particular field.

More information about this series at http://www.springer.com/series/8790

Gianluca Levi

Photoinduced Molecular Dynamics in Solution

Multiscale Modelling and the Link to Ultrafast Experiments

Doctoral Thesis accepted by
Technical University of Denmark,
Kongens Lyngby, Denmark

Author
Dr. Gianluca Levi
Science Institute of the University of Iceland
Reykjavík, Iceland

Supervisor
Prof. Klaus Braagaard Møller
Department of Chemistry
Technical University of Denmark (DTU)
Kongens Lyngby, Denmark

ISSN 2190-5053   ISSN 2190-5061 (electronic)
Springer Theses
ISBN 978-3-030-28613-2   ISBN 978-3-030-28611-8 (eBook)
https://doi.org/10.1007/978-3-030-28611-8

This Springer imprint is published by the registered company Springer Nature Switzerland AG
The registered company address is: Gewerbestrasse 11, 6330 Cham, Switzerland

to squeeze inside events,
dawdle in views,
to seek the least of all possible mistakes.
Wisława Szymborska
A note (Notatka)

Supervisor's Foreword

Mr. Levi's Ph.D. thesis concerns modelling photoinduced molecular dynamics in solution and establishing their link to ultrafast experiments. Considering that such experiments deliver information about the solution and the solvent in course of time, it is crucial to be able to reproduce the data by modelling it. This is key to understanding the reaction under study and come up with a rationale. The thesis tackles the case of a diplatinum complex in solution, which is a model system for photocatalysis. The thesis contains both experimental ultrafast X-ray scattering results obtained at LCLS in California as well as substantial computer simulations. For the latter purpose, Mr. Levi developed a multiscale protocol involving direct Born-Oppenheimer Molecular Dynamics using a QM/MM framework with the ΔSCF approach in the QM part. The computer simulation results were translated into X-ray scattering signals to provide direct contact with the experimental results. The thesis investigates the coherent vibrational dynamics in the ground and excited singlet states, providing deep insight into the energy flow within the molecule. The thesis work will inspire many researchers in the area of photoinduced solution dynamics, and the methods development will be useful for the theoretical chemistry community in general.

Kongens Lyngby, Denmark   Prof. Klaus Braagaard Møller
May 2019

Abstract

Recent advancements in X-ray source technologies have opened up the possibility for directly observing photoinduced chemical reactions as they unfold on the femtosecond time scale. An increasing number of time-resolved X-ray scattering experiments are being directed towards uncovering the light-induced ultrafast dynamics of photocatalytic metal complexes in solution. In this scenario, theory and modelling are brought into play to offer assistance to the interpretation and analysis of intricate measured data. Besides, theoretical modelling is the key to the fundamental understanding of the atomistic mechanisms behind reaction dynamics in solution.

The work presented in this thesis deals with extending, benchmarking and applying a novel multiscale atomistic modelling strategy for simulating the structural dynamics of complex molecular systems. The method is based on the direct Born-Oppenheimer Molecular Dynamics (BOMD) propagation of the nuclei and treats solvent effects within a quantum mechanics/molecular mechanics (QM/MM) framework.

The first part of the thesis shows how the QM/MM scheme is augmented to include electronic excited states with arbitrary spin multiplicity using a ΔSCF approach. We describe the testing and implementation of the method in the GPAW DFT code, providing all prerequisite theoretical background. The robustness of the implementation and the computational expediency of GPAW allow fast configurational sampling, overcoming the problem of statistical accuracy in excited-state BOMD simulations of systems as large as transition metal complexes.

The second part is dedicated to an investigation of the structure and dynamics of a model photocatalyst, the diplatinum(II) complex $[Pt_2(P_2O_5H_2)_4]^{4-}$, abbreviated PtPOP. In doing that we make extensive use of the computational tools presented in the first part. First, we show how ΔSCF for the first time provides computational evidence that the lowest-lying singlet and triplet excited states have parallel potential energy surfaces (PESs) along the Pt–Pt coordinate. Then we highlight the synergy between time-resolved experiments and simulations in unravelling the photoinduced ultrafast dynamics of the complex in water. QM/MM BOMD simulations are used to guide the analysis of X-ray diffuse scattering (XDS) data measured at an X-ray free-electron laser (XFEL), and to elaborate a semi-classical picture of ground-state hole dynamics that explains the experimental outcome. Finally, we take a step forward in the understanding of the excited-state vibrational relaxation in solution. We show, through the simulations, that PtPOP after excitation does not retain the symmetry of the ground state, as so far believed, and that excess Pt–Pt vibrational energy is first directed towards vibrational modes involving the ligands, while the role of the solvent is to favour intramolecular vibrational energy redistribution (IVR) in the complex.

Preface

This thesis has been submitted to the Department of Chemistry, Technical University of Denmark, in partial fulfilment of the requirements for the Ph.D. degree in the subject of chemistry. The work presented herein was carried out at the Department of Chemistry, Technical University of Denmark, from December 2014 to January 2018, under the supervision of Prof. Klaus B. Møller, and joint co-supervision of Postdoc Asmus O. Dohn and Assoc. Prof. Niels E. Henriksen. In addition, part of the work was carried out during a one-week experimental campaign at the LCLS XFEL facility of Stanford between February and March 2015, and at the Faculty of Physical Sciences, University of Iceland, in the group of Prof. Hannes Jónnson in March 2017.

Kongens Lyngby, Denmark
February 2018

Dr. Gianluca Levi

Parts of this thesis have been published in the following journal articles:

The work presented in this thesis resulted in scientific contributions in terms of theoretical and code development, and advancements in the understanding of the PtPOP system. Below, we list the articles that were published based on the main achievements of the thesis work.

Theoretical work:

- Contribution to the theoretical formulation of the GPAW electrostatic embedding QM/MM scheme [1].

Computational development:

- Implementation of a ΔSCF scheme with Gaussian smeared constraints in the GPAW DFT code (https://gitlab.com/glevi/gpaw/tree/Dscf_gauss) [2].
- Implementation of a classical counterion model with spherical harmonic restraints within the TIP4P force field originally implemented by A. O. Dohn for ASE (developed in-house) [1, 2].
- Improvements to the ASE modules for calculating Lennard-Jones interactions and applying RATTLE constraints in MD simulations (https://gitlab.com/glevi).

Contributions to the understanding of the photophysics, dynamics and solution properties of PtPOP:

- Theoretical and computational support to the first experimental determination of the change in Pt–Pt equilibrium distance in the lowest-lying singlet excited state in water by X-ray diffuse scattering measurements [3].
- First time computational proof of the similarities between the potential energy surfaces in the lowest singlet and triplet excited states [2].
- QM/MM characterization of the solvation shell in water in the ground and first singlet excited states [1, 2].
- First time computational evidence of ligand distortion and symmetry breaking in the excited state [2].
- Elaboration of a semi-classical model to interpret the experimental observation of ground-state dynamics, and time-resolved XDS determination of the Pt–Pt vibrational period of the ground state [4].
- Computational identification of the paths for vibrational relaxation in the first singlet excited state in water [2].

[1] A. O. Dohn, E. Ö. Jónsson, **G. Levi**, J. J. Mortensen, O. Lopez-Acevedo, K. S. Thygesen, K. W. Jacobsen, J. Ulstrup, N. E. Henriksen, K. B. Møller, and H. Jónsson. Grid-Based Projector Augmented Wave (GPAW) implementation of Quantum Mechanics/Molecular Mechanics (QM/MM) electrostatic embedding and application to a solvated diplatinum complex. *Journal of Chemical Theory and Computation*, 13(12):6010–6022, 2017.

[2] **G. Levi**, M. Pápai, N. E. Henriksen, A. O. Dohn, and K. B. Møller. Solution structure and ultrafast vibrational relaxation of the PtPOP complex revealed by ΔSCF-QM/MM Direct Dynamics simulations. *Journal of Physical Chemistry C*, 122:7100–7119, 2018.

[3] E. Biasin, T. B. van Driel, **G. Levi**, M. G. Laursen, A. O. Dohn, A. Moltke, P. Vester, F. B. K. Hansen, K. S. Kjaer, R. Hartsock, M. Christensen, K. J. Gaffney, N. E. Henriksen, K. B. Møller, K. Haldrup, and M. M. Nielsen. Anisotropy enhanced X-ray scattering from solvated transition metal complexes. *Journal of Synchrotron Radiation*, 25(2):306–315, 2018.

[4] K. Haldrup, **G. Levi**, E. Biasin, P. Vester, M. G. Laursen, F. Beyer, K. S. Kjær, T. Brandt van Driel, T. Harlang, A. O. Dohn, R. J. Hartsock, S. Nelson, J. M. Glownia, H. T. Lemke, M. Christensen, K. J. Gaffney, N. E. Henriksen, K. B. Møller, and M. M. Nielsen. Ultrafast x-ray scattering measurements of coherent structural dynamics on the ground-state potential energy surface of a diplatinum molecule. *Phys. Rev. Lett.*, 122:063001, 2019.

Acknowledgements

First of all, let me express my sincere gratitude to my three supervisors, Klaus B. Møller, Asmus O. Dohn and Niels E. Henriksen. Throughout this Ph.D. project, I could draw upon their expertise and experience in the field, while receiving constant support and encouragement. However, my appreciation is not limited to that. Klaus, your astonishing capacity for looking deeper into all aspects of the scientific work (and not only that) has been for me source of great inspiration and personal and professional development. I don't think there has been a single meeting with you in which I haven't learnt something about how to "seek the least of all possible mistakes" or how to look at things from a broader perspective. Asmus, the dedication you showed to the supervision of your first Ph.D. student is remarkable. During these three years, I could always count on your help, guidance and advice. Of course, thanks for having launched me with such enthusiasm on the QM/MM project in ASE, for having introduced me to several aspects of Danish and Icelandic cultures, and shown the tricks of the trade in the life of a Ph.D. student in Denmark (and the list could continue, but I will stop here because I know you don't like to be thanked too much). Niels, thanks a lot for all the time you spent supplying me with precious indications for my research. The modelling and understanding of the ground-state dynamics of PtPOP have greatly benefited from all information and assistance I received from you.

I would like to thank the whole group of X-ray scientists lead by Martin M. Nielsen at DTU Physics, with whom I collaborated. Thanks Martin for setting the stage for insightful discussions in a very nice atmosphere during our group meetings. I am greatly indebted to Kristoffer Haldrup for coordinating the on-site analysis in which I was involved during the PtPOP experiments at LCLC, at a time when I was still green in Matlab coding and X-ray scattering. Your help, and showing around the beamline, is very much appreciated! Thanks also to Frederik B. K. Hansen for having been always ready to help debugging during the PtPOP beamtime. The role of Elisa Biasin in anchoring the PtPOP simulations to the experiments has been invaluable. Your fantastic work in the analysis of the experimental data has been fundamental for this project, Elisa. Thanks also for all your encouragement and insightful discussions. The work done by Peter Vester in the analysis of the PtPOP data is also acknowledged. Still among our X-ray collaborators, let me thank Mads Laursen and Diana B. Zederkof for insights into the experiments and good times at conferences. Finally, I would like to extend my thanks to those I consider the rock stars of X-ray science: Kasper S. Kjær, Tim B. van Driel, Tobias C. B. Harlang and Morten Christensen, whose technical skills in the field are only exceeded by their extraordinary enthusiasm towards ultrafast discoveries. Without them the experimental campaign on PtPOP would have never been possible.

Let me express my thanks to Hannes Jónsson, who welcomed me for one month in his group at University of Iceland, in a highly stimulating environment, and to Elvar Ö. Jónsson, always ready to clarify my questions about the QM/MM code.

Doing computational work means also sharing an office for many hours a day, for many days in a year, with the same persons. Thus, I take the opportunity to thank my office colleagues at DTU Chemistry, Mats Simmermacher, Mátyás Pápai, Mostafa Abedi and Esben F. Thomas, for having created a highly peculiar environment where it's difficult to get bored (at least if you like nerdy math jokes and are reticent to whatsoever conventional form of sociality). Thanks for always keeping up the good humour guys. Mats, thanks for the countless discussions on the most various topics, for being always willing to share your profound knowledge of quantum mechanics with great enthusiasm, and for reminding me about my Neapolitan origins (yes, I am thinking about "la distribuzione di Napoli"). I hope I will be able to repay you by visiting Lübeck, finally. Mátyás, I am very grateful to you for the numberless DFT calculations on PtPOP, and for always answering my often basic questions on electronic structure theory. Of course, thanking you also for all the pálinka is imperative. Many thanks to Mostafa for insightful discussions about classical MD models, and to Esben for helping me in the understanding of different mathematical problems, and for stimulating my brain with other, more involved ones.

For a combination of factors, the last months of this Ph.D. have been surprisingly hectic. During this period, I could benefit greatly from the support and encouragement of Alexandre, Cécile, Luigi, Marta, Peter and Stefan (and probably others, I am sorry if I have forgotten you!). Thanks a lot, guys.

This Ph.D. project would have never been possible without the support of my family. Let me thank them. Mamma, grazie per tutto l'incoraggiamento negli ultimi mesi. Sono consapevole che non è facile sentire il proprio figlio lontano e apparentemente slegato dal suo luogo di origine; ma sappi che le proprie radici si conservano sempre, indubbiamente. Padre, grazie per gli insegnamenti di matematica da quando facevi lezione con i lucidi e il proiettore fino al Demidovic, affidatomi prima di partire. Questo ha contribuito a maturare in me, se non una profonda passione, una necessaria curiosità per la ricerca scientifica. Forse vi perdono per avermi scoraggiato a seguire il sentiero umanistico.

Lastly, my special thanks go to Julia who has been brave enough to accompany me alongside this journey. Ju, no word is enough to express how much I owe you for all the support I received throughout these years. I just want you to know that the equations, funky plots and snapshots of atomic motion that are contained in this thesis are inextricably linked to you, as they are in a way the result of our endless walks in city and in forest.

Contents

Abbreviations

AIMS	Ab Initio Multiple Spawning
ASE	Atomic Simulation Environment
BO	Born-Oppenheimer
BOMD	Born-Oppenheimer Molecular Dynamics
DFT	Density Functional Theory
DOF	Degrees of Freedom
DOS	Density of States
ECP	Effective Core Potential
ES	Excited State
FT	Fourier Transform
FWHM	Full Width at Half Maximum
GGA	Generalized Gradient Approximation
GPAW	Grid-based Projector Augmented Wave
GS	Ground State
GTO	Gaussian-Type Orbitals
HF	Hartree-Fock
HOMO	Highest Occupied Molecular Orbital
IAM	Independent Atom Model
IRF	Instrument Response Function
ISC	Intersystem Crossing
IVR	Intramolecular Vibrational Energy Redistribution
KS	Kohn-Sham
LCAO	Linear Combination of Atomic Orbitals
LCLS	Linac Coherent Light Source
LDA	Local Density Approximation
leΔSCF	linear expansion Δ-Self-Consistent Field
LJ	Lennard-Jones
LUMO	Lowest Unoccupied Molecular Orbital
MD	Molecular Dynamics
MM	Molecular Mechanics

MOM	Maximum Overlap Method
PAW	Projector Augmented Wave
PBC	Periodic Boundary Condition
PES	Potential Energy Surface
PMF	Potential of Mean Force
PP	Pseudo-Potential
QM	Quantum Mechanics
QM/MM	Quantum Mechanics/Molecular Mechanics
RCS	Restricted Closed-Shell
RDF	Radial Distribution Function
RISRS	Resonant Impulsive Stimulated Raman Scattering
RMSD	Root-Mean-Square Deviation
RMSE	Root-Mean-Square Error
ROS	Restricted Open-Shell
SCF	Self-Consistent Field
ΔSCF	Δ-Self-Consistent Field
SF	Spatial Filtering
SOC	Spin-Orbit Coupling
SVD	Singular-Value Decomposition
TDDFT	Time-Dependent Density Functional Theory
TSH	Trajectory Surface Hopping
UOS	Unrestricted Open-Shell
vdW	van der Waals
XDS	X-ray Diffuse Scattering
XFEL	X-ray Free-Electron Laser
XPP	X-ray Pump Probe

List of Figures

List of Tables

Part I
Introduction and Background

Chapter 1
Filming Motion at the Atomic Scale of Time

Being able to observe the dynamics of the chemical bond in real time has been one of the greatest achievements of modern physical chemistry over the last three decades. Before then, the motion of atoms during bond-breaking/forming reactions had been inaccessible to direct experimental observation. The reason lies in the ultrafast nature of these atomistic processes. Indeed, nuclear vibrational motion unfolds on a very short time scale, the femtosecond time scale (1 fs $= 10^{-15}$ s). Femtochemistry [2, 3], the study of reaction intermediates at the atomic scale of time, started out with the ultrafast experiments performed by A. H. Zewail in the late 1980s, for which he was awarded the 1999 Nobel Prize in Chemistry [4]. The pioneering experiments investigated the dissociation of diatomic [5] and triatomic [6, 7] molecules in gas phase, and were made possible by the advent of ultrashort optical laser technologies. Clocking of such ultrafast chemical processes is achieved according to the pump-probe methodology. A femtosecond optical pulse is used to initiate the coherent and synchronous motion of the atoms. This first pump pulse is followed, after a time delay controlled with femtosecond resolution, by a second ultrashort pulse of radiation, the probe pulse, which captures an individual snapshot of atomic motion. Combining snapshots recorded in a sequence of pump-probe time delays produces a "motion picture" of the dynamics. Since nuclear dynamics is an intrinsic reflection of the reaction mechanisms, pump-probe technologies have paved the way to the mechanistic understanding of an increasingly ample range of chemical reactions.

All early pump-probe investigations employed an optical UV-vis probe. However, spectroscopic data do not correlate directly to structural changes. The structural information can be inferred indirectly from optical measurements if detailed knowledge of the electronic structure of the system is available. While this can be true in the case of elementary reactions involving small diatomic and triatomic molecules, as the system grows in size, extensive electronic structure calculations are needed, which

---

Parts of this chapter have been reproduced with permission from Ref. [1], https://doi.org/10.1021/acs.jpcc.8b00301. Copyright 2018 American Chemical Society.

G. Levi, *Photoinduced Molecular Dynamics in Solution*, Springer Theses, https://doi.org/10.1007/978-3-030-28611-8_1

3

can be cumbersome at best. The complexity of the problem is particularly high when dealing with molecular reactions in solution, since the degrees of freedom involved are many and the dynamics is inherently dominated by time-dependent distributions of atomic positions. On the other hand, X-rays can provide a more direct probe of the photoinduced structural changes. This was understood soon after the first optical pump-probe experiments [8]. The challenge to proceed further along this direction has been represented by the design and implementation of coherent X-ray sources capable of providing femtosecond time resolution and sufficiently high photon flux. Nowadays, novel X-ray free-electron laser (XFEL) facilities [8–11] meet all the requisites needed to image atomic motion in solution with X-rays.

1.1 Ultrafast Studies of Transition Metal Complexes

Photocatalytic reactions involving transition metal complexes in solution have been among the most popular targets of time-resolved experiments over the last years [8, 12, 13]. Indeed, stability in solution, remarkable photophysical properties and the presence of electron-rich atoms, make transition metal complexes attractive candidates for both spectroscopic and X-ray ultrafast studies. Taking full advantage of their photocatalytic properties requires an understanding of the structure-function relationships and mechanisms behind ultrafast light-induced reactions in complex environments. The continuous demand for more efficient photocatalitic systems combined with tremendous advancements in pump-probe techniques has led to a whole host of experiments able to follow the evolution of vibrational wave packets or the solvation dynamics in photoexcited prototypical metal complexes in real time [14–21].

These novel experiments cover grounds often dominated by complex interplays between vibrational relaxation, solvent effects and electronic couplings, which are not known a priori. Therefore, linking experimental observations to mechanistic frameworks can only be accomplished with the help of solid theoretical and modelling strategies. Moreover, even when the interpretation of an experiment is facilitated by prior photophysical knowledge or by employing simple phenomenological models, a variety of complementary techniques are needed to assemble a complete atomistic and energetic picture of the early stages of the investigated dynamics. In this context, advanced computational methods capable of connecting multiple time-resolved observables, while delivering new mechanistic insights into the underlying physical processes, play an important role in complementing ultrafast experiments of transition metal complexes.

1.2 Modelling Strategies

One of the main challenges associated with ab initio computational determination of the mechanisms of the ultrafast excited-state dynamics of complex molecular systems is represented by the time scales one is able to simulate while retaining accuracy. As experimental techniques with atomistic resolution start putting a lens onto hitherto unexplored sub-picosecond intramolecular structural and solvation processes, developing efficient computational methods capable of providing insights into the underlying physical mechanisms becomes of utmost importance. Broadly speaking, much of the efforts of the theoretical community to address this challenge have been directed towards the development and application of two computational frameworks of choice: methods that solve the time-dependent Schrödinger equation for the nuclei using precomputed potential energy surfaces (PESs) [22–28], and methods based on classical propagation of the nuclei with on-the-fly evaluation of energies and forces at ab initio level [29–34]. Quantum dynamics approaches have proven useful in deciphering some aspects of the excited-state decay pathways of photocatalytic metal complexes, particularly concerning non-adiabatic electronic transitions [22, 25]. However, the outcome of this kind of simulation relies on the selection of a small number of vibrational modes along which the dynamics is restricted. Furthermore, solvent effects in quantum wave packet simulations are usually accounted for in an implicit manner [24, 26], thus neglecting any explicit solvation dynamics effect. On the other hand, the second approach, ab initio classical dynamics, allows, in principle, to efficiently explore the full, unconstrained space of nuclear configurations and to include explicit solvent effects in a multiscale fashion. The price to pay for having abandoned a quantum description of the dynamics, is that quantum effects, like non-adiabatic electronic transitions and tunnelling, are neglected in this second picture. In particular, neglecting the non-adiabatic couplings between electronic and nuclear motions implies restricting the dynamics of the nuclei to a single, Born-Oppenheimer (BO) PES (the concept of BO PES will be introduced in Chap. 4). Cases in which non-adiabatic effects are important on the time scales that are considered in the investigation, can be treated, without abandoning the full-dimensionality provided by the classical trajectory description, with mixed quantum-classical methods like trajectory surface hopping (TSH) [35–37], or the closely related ab initio multiple spawning (AIMS) [38–40]. The basic idea behind these approaches is that the time evolution of a non-adiabatic system can be reproduced by ensembles of trajectories that evolve on BO electronic surfaces and experience state switches in proximity of regions of non-adiabaticity. Among them, TSH has been the one that has been most extensively applied to study the mechanisms behind the first steps of the ultrafast relaxation cascade of photoexcited metal complexes [41–44].

The work presented in this thesis focused principally on extending the features and capabilities of, and applying a multiscale computational method [45–47] that follows along the second modelling strategy. The approach is based on a density functional theory (DFT) implementation of on-the-fly quantum mechanics/molecular mechanics (QM/MM) Born-Oppenheimer Molecular Dynamics (BOMD) [48]. The implementation is available within the Atomic Simulation Environment (ASE) [49, 50] and uses the computationally efficient Grid-based Projector Augmented Wave (GPAW) code [51, 52] for the DFT part. In its basic form, it was already available before the start of the present Ph.D. project, and had already been successfully applied to study the ultrafast internal vibrational dynamics and to obtain a picture of solvent-driven electronic dynamics in bimetallic photoactive complexes [31, 53]. More specifically, the method is tailored to help the interpretation and analysis of optical pump-X-ray probe experiments on transition metal complexes in solution. The experiments are performed by the group where the Ph.D. project took place together with experimental collaborators at XFELs facilities. As we will see in more detail in the course of the thesis, X-ray scattering signals of solvated molecules are much more challenging to analyse than conventional X-ray scattering patterns of crystals. Put simply, the X-ray scattering of a solution appears diffuse (and for this reason it is referred to as "X-ray diffuse scattering" (XDS)), lacking the characteristic Bragg peaks of the scattering signal of periodic systems, which allow to infer directly structural information. Our multiscale approach offers support to the characterization of time-resolved XDS data by delivering statistically relevant and accurate information on both thermal equilibrium properties and ultrafast out-of-equilibrium dynamical processes. For example, the method has proven decisive in establishing a robust interpretation of the solvation dynamics at the catalytic site of a diiridium complex observed in ultrafast XDS data [17]. We have recently presented the full details of the QM/MM BOMD implementation in ASE and GPAW in Ref. [48].

In all previous applications, the excited states of interest were described using the spin unrestricted DFT formalism. In some of the investigated systems, the observed ultrafast dynamics following photoexcitation was known to take place on an excited state of the same spin multiplicity as the ground state, usually a singlet. This implied that the simulations had to approximate the dynamics by propagating the system on the lowest excited state of a different spin multiplicity by assuming parallel PESs along the dominant vibrational motions. However, even in systems for which the latter assumption was demonstrated to be valid, the dynamics in the two states can still be different if their energies are such that they lie in regions of different density of states, as recently shown by Monni et al. [15]. These authors compared the coherence decay of vibrational wave packets in the first singlet and triplet excited states of diplatinum complexes in solution observed in ultrafast optical measurements, and found significant differences despite parallel PESs.

The need to be able to reliably compare simulations to experimental results calls for an extension of the QM/MM BOMD method in ASE/GPAW to encompass states of arbitrary spin multiplicity. With this perspective in mind, part of the work carried out during the present Ph.D. project [1] has been devoted to extending the capabilities of the code by coupling it to a single-determinant DFT description of the excited states based on the Δ-self-consistent-field (ΔSCF) approach [54], which carries no extra computational cost with respect to ground-state DFT.

ΔSCF is gaining increasing popularity in the study of the excited states of both organic chromophores [55–58] and transition metal complexes [59, 60]. This renewed interest is motivated in part by the growing demand for computationally cheap strategies for simulating with sufficient accuracy the excited-state structure and dynamics of large systems, for which high-level multireference methods are not yet a viable choice. The reliability of ΔSCF as applied to study the structure and dynamics of small molecules, organic dyes and even biological systems, has been assessed with respect to vibrational analysis [61], exploration of PESs [58, 62], as well as dynamics in solution within QM/MM MD frameworks [56, 63]. On the other hand, to our knowledge, no studies exist that investigate the ability of the method to predict the structural dynamics of transition metal complexes, even though the performances of ΔSCF for excitation energies and simulations of UV-vis spectra of metal-containing molecular systems are not inferior to those achieved when applied to organic molecules [59, 60]. A second general objective of the present work has been to assess the reliability of ΔSCF for prediction of structural and dynamical properties of transition metal complexes.

We note that the understanding of the processes that govern the ultrafast excited-state dynamics of transition metal complexes has greatly benefited from simulations using other MD codes. Among them, the ones that have gained most popularity for the study of transition metal complexes are probably the SHARC program [29, 35, 41, 64] and the plane-wave code CPMD [33, 42, 44, 65–67]. These software packages are quite advanced, they include interfaces to a host of electronic structure codes, as in the case of SHARC, can work with QM/MM schemes, and implement non-adiabatic MD in a surface hopping perspective. On the other hand, they have all employed DFT in its time-dependent formulation (TDDFT) to describe the excited states of transition metal complexes. Our implementation of excited-state QM/MM BOMD is, instead, unique in its combination of a cost-effective single determinant method as ΔSCF with the computationally expedient GPAW DFT code. Therefore, we see our ΔSCF-QM/MM BOMD method not as a step back with respect to already existing MD codes, but rather as a complementary technique, which can turn especially useful when statistical significance and an explicit description of solvation effects can be privileged over, for example, the inclusion of non-adiabatic effects, as we will see throughout this thesis.

References

1. Levi G, Pápai M, Henriksen NE, Dohn AO, Møller KB (2018) Solution structure and ultrafast vibrational relaxation of the PtPOP complex revealed by ΔSCF-QM/MM direct dynamics simulations. J Phys Chem C 122:7100–7119
2. Henriksen NE (2014) Femtochemistry—some reflections and perspectives. Chem Phys 442:2–8
3. Zewail AH (2000) Femtochemistry: atomic-scale dynamics of the chemical bond using ultrafast lasers (nobel lecture). Angew Chem Int Ed 39(15):2586–2631
4. The Nobel Prize in Chemistry (1999). https://www.nobelprize.org/nobel_prizes/chemistry/laureates/1999/. Accessed 04 February 2018
5. Rose TS, Rosker MJ, Zewail AH (1988) Femtosecond real-time observation of wave packet oscillations (resonance) in dissociation reactions. J Chem Phys 88:6672

6. Dantus M, Rosker MJ, Zewail AH (1987) Real-time femtosecond probing of "transition states" in chemical reactions. J Chem Phys 87:2395
7. Scherer NF, Knee JL, Smith DD, Zewail AH (1985) Femtosecond photofragment spectroscopy of the reaction iodine cyanide (ICN) → cyanogen (CN) + atomic iodine. J Phys Chem 89:5141–5143
8. Chergui M, Collet E (2017) Photoinduced structural dynamics of molecular systems mapped by time-resolved x-ray methods. Chem Rev 117(16):11025–11065
9. Bostedt C, Boutet S, Fritz DM, Huang Z, Lee HJ, Lemke HT, Robert A, Schlotter WF, Turner JJ, Williams GJ Linac (2016) Coherent light source: the first five years. Rev Modern Phys 88(1):015007
10. Chollet M, Alonso-Mori R, Cammarata M, Damiani D, Defever J, Delor JT, Feng Y, Glownia JM, Langton JB, Nelson S, Ramsey K, Robert A, Sikorski M, Song S, Stefanescu D, Srinivasan V, Zhu D, Lemke HT, Fritz DM (2015) The x-ray pump-probe instrument at the linac coherent light source. J Synchrotron Radiat 22:503–507
11. Emma P, Akre R, Arthur J, Bionta R, Bostedt C, Bozek J, Brachmann A, Bucksbaum P, Coffee R, Decker FJ, Ding Y, Dowell D, Edstrom S, Fisher A, Frisch J, Gilevich S, Hastings J, Hays G, Ph Hering Z, Huang R, Iverson H, Loos M, Messerschmidt A, Miahnahri S, Moeller HD, Nuhn G, Pile D, Ratner J, Rzepiela D, Schultz T, Smith P, Stefan H, Tompkins J, Turner J, Welch J, White W, Wu J, Tocky G, Galayda J (2010) First lasing and operation of an ångstrom-wavelength free-electron laser. Nat Photonics 4(9):641–647
12. Gray HB, Záliš S, Vlček A (2017) Electronic structures and photophysics of d8–d8 complexes. Coordination Chem Rev 345:297–317
13. Chergui M (2015) Ultrafast photophysics of transition metal complexes. Acc Chem Res 48:801–808
14. Lemke HT, Kjær KS, Hartsock R, Van Driel TB, Chollet M, Glownia JM, Song S, Zhu D, Pace E, Matar SF, Nielsen MM, Benfatto M, Gaffney KJ, Collet E, Cammarata M (2017) Coherent structural trapping through wave packet dispersion during photoinduced spin state switching. Nat Commun 8(May):15342
15. Monni R, Auböck G, Kinschel D, Aziz-Lange KM, Gray HB, Vlček A, Chergui M (2017) Conservation of vibrational coherence in ultrafast electronic relaxation: the case of diplatinum complexes in solution. Chem Phys Lett 683:112–120
16. Biasin E, van Driel TB, Kjær KS, Dohn AO, Christensen M, Harlang T, Chabera P, Liu Y, Uhlig J, Pápai M, Németh Z, Hartsock R, Liang W, Zhang J, Alonso-Mori R, Chollet M, Glownia JM, Nelson S, Sokaras D, Assefa TA, Britz A, Galler A, Gawelda W, Bressler C, Gaffney KJ, Lemke HT, Møller KB, Nielsen MM, Sundström V, Vankó G, Wärnmark K, Canton SE, Haldrup K (2016) Femtosecond x-ray scattering study of ultrafast photoinduced structural dynamics in solvated [Co(terpy)2]2+. Phys Rev Lett 117(1):013002
17. van Driel TB, Kjær KS, Hartsock R, Dohn AO, Harlang T, Chollet M, Christensen M, Gawelda W, Henriksen NE, Kim JG, Haldrup K, Kim KH, Ihee H, Kim J, Lemke H, Sun Z, Sundstrom V, Zhang W, Zhu D, Møller KB, Nielsen MM, Gaffney KJ (2016) Atomistic characterization of the active-site solvation dynamics of a photocatalyst. Nat Commun 7:13678
18. Hua L, Iwamura M, Takeuchi S, Tahara T (2015) The substituent effect on the MLCT excited state dynamics of Cu(I) complexes studied by femtosecond time-resolved absorption and observation of coherent nuclear wavepacket motion. Phys Chem Chem Phys 17(3):2067–2077
19. van der Veen RM, Cannizzo A, van Mourik F, Vlček Jr A, Chergui M (2011) Vibrational relaxation and intersystem crossing of binuclear metal complexes in solution. J Am Chem Soc 113:305
20. Iwamura M, Watanabe H, Ishii K, Takeuchi S, Tahara T (2011) Coherent nuclear dynamics in ultrafast photoinduced structural change of Bis(diimine)copper(I) complex. J Am Chem Soc 133(20):7728–7736
21. Consani C, Prémont-Schwarz M, Elnahhas A, Bressler C, Van Mourik F, Cannizzo A, Chergui M (2009) Vibrational coherences and relaxation in the high-spin state of aqueous [FeII(bpy)3]2+. Angew Chemie Int Edition 48(39):7184–7187

22. Fumanal M, Gindensperger E, Daniel C (2017) Ultrafast excited-state decays in [Re(CO)3(N, N)(L)]n+: nonadiabatic quantum dynamics. J Chem Theory Comput 13(3):1293–1306
23. Pápai M, Penfold TJ, Møller KB (2016) Effect of tert-butyl functionalization on the Photoexcited Decay of a Fe(II)-N-Heterocyclic carbene complex. J Phys Chem C 120(31):17234–17241
24. Harabuchi Y, Eng J, Gindensperger E, Taketsugu T, Maeda S, Daniel C (2016) Exploring the mechanism of ultrafast intersystem crossing in rhenium(I) carbonyl bipyridine halide complexes: key vibrational modes and spin-vibronic quantum dynamics. J Chem Theory Comput 12(5):2335–2345
25. Pápai M, Vankó G, Rozgonyi T, Penfold TJ (2016) High efficiency iron photosensitiser explained with quantum wavepacket dynamics. J Phys Chem Lett 7:2009–2014
26. Eng J, Gourlaouen C, Gindensperger E, Daniel C (2015) Spin-vibronic quantum dynamics for ultrafast excited-state processes. Acc Chem Res 48(3):809–817
27. Beck MH, Jäckle A, Worth GA, Meyer H-D (1999) The multiconfiguration time-dependent Hartree (MCTDH) method: a highly efficient algorithm for propagating wavepackets. Phys Rep 324:1–105
28. Meyer HD, Manthe U, Cederbaum LS (1990) The multi-configurational time-dependent Hartree approach. Chem Phys Lett 165(1):73–78
29. Mai S, Gattuso H, Fumanal M, Muñoz-Losa A, Monari A, Daniel C, Gonzalez L (2017) Excited-states of a rhenium carbonyl diimine complex: solvation models, spin-orbit coupling, and vibrational sampling effects. Phys Chem Chem Phys 3:21–25
30. Dohn AO, Henriksen NE, Møller KB (2014) Transient changes in molecular geometries and how to model them. Springer International Publishing, 2014
31. Dohn AO, Jónsson EÖ, Kjær KS, van Driel TB, Nielsen MM, Jacobsen KW, Henriksen NE, Møller KB (2014) Direct dynamics studies of a binuclear metal complex in solution: the interplay between vibrational relaxation, coherence, and solvent effects. J Phys Chem Lett 5:2414–2418
32. Daku LML, Hauser A (2010) Ab initio molecular dynamics study of an aqueous solution of [Fe(bpy)3](Cl)2 in the low-spin and in the high-spin states. J Phys Chem Lett 1:1830–1835
33. Moret M-E, Tavernelli I, Rothlisberger U (2009) Combined QM/MM and classical molecular dynamics study of [Ru(bpy)3]2+ in water, 113:7737–7744
34. Marx D, Hutter J (2009) Ab initio molecular dynamics: basic theory and advanced methods. Cambridge University Press
35. Mai S, Marquetand P, González L (2015) A general method to describe intersystem crossing dynamics in trajectory surface hopping. Int J Quantum Chem 115:1215–1231
36. Malhado JP, Bearpark MJ, Hynes JT (2014) Non-adiabatic dynamics close to conical intersections and the surface hopping perspective. Frontiers Chem 2(97):1–21
37. Tully JC (1990) Molecular dynamics with electronic transitions. J Chem Phys 93(2):1061
38. Curchod BFE, Rauer C, Marquetand P, González L, Martínez TJ (2016) Communication: GAIMS-generalized ab initio multiple spawning for both internal conversion and intersystem crossing processes. J Chem Phys 144(10):101102
39. Ben-Nun M, Martínez TJ (2002) Ab initio quantum molecular dynamics, pp 439–512. John Wiley & Sons, Inc., 2002
40. Ben-Nun M, Quenneville J, Martínez TJ (2000) Ab initio multiple spawning: photochemistry from first principles quantum molecular dynamics. J Phys Chem A 104(22):5161–5175
41. Atkins AJ, González L (2017) Trajectory surface-hopping dynamics including intersystem crossing in [Ru(bpy)3]2+. J Phys Chem Lett 8(16):3840–3845
42. Capano G, Penfold TJ, Chergui M, Tavernelli I (2017) Photophysics of a copper phenanthroline elucidated by trajectory and wavepacket-based quantum dynamics: a synergetic approach. Phys Chem Chem Phys 19:19590–19600
43. Tavernelli I (2015) Nonadiabatic molecular dynamics simulations: synergies between theory and experiments. Acc Chem Res 48:792–800
44. Tavernelli I, Curchod BFE, Rothlisberger U (2011) Nonadiabatic molecular dynamics with solvent effects: a LR-TDDFT QM/MM study of ruthenium (II) tris (bipyridine) in water. Chem Phys 391:101–109

45. The Nobel Prize in Chemistry (2013), https://www.nobelprize.org/nobel_prizes/chemistry/laureates/2013/. Accessed 11 February 2018
46. Field MJ, Bash PA, Karplus M (1990) A combined quantum mechanical and molecular mechanical potential for molecular dynamics simulations. J Comput Chem 11(6):700–733
47. Warshel A, Levitt M (1976) Theoretical studies of enzymatic reactions: dielectric, electrostatic and steric stabilization of the carbonium ion in the reaction of lysozyme. J Mol Biol 103:227–249
48. Dohn AO, Jónsson EÖ, Levi G, Mortensen JJ, Lopez-Acevedo O, Thygesen KS, Jacobsen KW, Ulstrup J, Henriksen NE, Møller KB, Jónsson H (2017) Grid-based projector augmented wave (GPAW) implementation of quantum mechanics/molecular mechanics (QM/MM) electrostatic embedding and application to a solvated diplatinum complex. J Chem Theory Comput 13(12):6010–6022
49. Larsen AH, Mortensen JJ, Blomqvist J, Castelli IE, Christensen R, Dułak M, Friis J, Groves MN, Hammer B, Hargus C, Hermes ED, Jennings PC, Jensen PB, Kermode J, Kitchin JR, Kolsbjerg EL, Kubal J, Kaasbjerg K, Lysgaard S, Maronsson JB, Maxson T, Olsen T, Pastewka L, Peterson A, Rostgaard C, Schiøtz J, Schütt O, Strange M, Thygesen KS, Vegge T, Vilhelmsen L, Walter M, Zeng Z, Jacobsen KW (2017) The atomic simulation environment-a python library for working with atoms. J Phys Condens Matter 29(27):273002
50. Bahn SR, Jacobsen KW (2002) An object-oriented scripting interface to a legacy electronic structure code. Comput Sci Eng 4:55
51. Enkovaara J, Rostgaard C, Mortensen JJ, Chen J, Dulak M, Ferrighi L, Gavnholt J, Glinsvad C, Haikola V, Hansen HA, Kristoffersen HH, Kuisma M, Larsen AH, Lehtovaara L, Ljungberg M, Lopez-Acevedo O, Moses PG, Ojanen J, Olsen T, Petzold V, Romero NA, Stausholm-Møller J, Strange M, Tritsaris GA, Vanin M, Walter M, Hammer B, Häkkinen H, Madsen GKH, Nieminen RM, Nørskov JK, Puska M, Rantala TT, Schiøtz J, Thygesen KS, Jacobsen KW (2010) Electronic structure calculations with GPAW: a real-space implementation of the projector augmented-wave method. J Phys Condens Matter 22:253202
52. Mortensen JJ, Hansen LB, Jacobsen KW (2005) Real-space grid implementation of the projector augmented wave method. Phys Rev B 71:035109
53. Dohn AO, Kjær KS, Harlang TB, Canton SE, Nielsen MM, Møller KB (2016) Electron transfer and solvent-mediated electronic localization in molecular photocatalysis. Inorg Chem 55(20):10637–10644
54. Ziegler T, Rauk A, Baerends EJ (1977) On the calculation of multiplet energies by the hartree-fock-slater method. Theor Chim Acta 43(3):261–271
55. Briggs EA, Besley NA (2015) Density functional theory based analysis of photoinduced electron transfer in a triazacryptand based K+ sensor. J Phys Chem A 119:2902–2907
56. Briggs EA, Besley NA, Robinson D (2013) QM/MM excited state molecular dynamics and fluorescence spectroscopy of BODIPY. J Phys Chem A 117(12):2644–2650
57. Kowalczyk T, Yost SR, Van Voorhis T (2011) Assessment of the ΔSCF density functional theory approach for electronic excitations in organic dyes. J Chem Phys 134(5):054128
58. Maurer RJ, Reuter K (2011) Assessing computationally efficient isomerization dynamics: ΔSCF density-functional theory study of azobenzene molecular switching. J Chem Phys 135(22):1–25
59. Himmetoglu B, Marchenko A, Dabo I, Cococcioni M (2012) Role of electronic localization in the phosphorescence of iridium sensitizing dyes. J Chem Phys 137(15):154309
60. Robinson D, Besley NA (2010) Modelling the spectroscopy and dynamics of plastocyanin. Phys Chem Chem Phys 12(33):9667–9676
61. Hanson-Heine MWD, George MW, Besley NA (2013) Calculating excited state properties using Kohn-Sham density functional theory. J Chem Phys 138(6):064101
62. Gavnholt J, Olsen T, Engelund M, Schiøtz J (2008) Delta self-consistent field as a method to obtain potential energy surfaces of excited molecules on surfaces. Phys Rev B 78:075441
63. Mendieta-Moreno J, Trabada DG, Mendieta J, Lewis JP, Gómez-Puertas P, Ortega J (2016) Quantum mechanics/molecular mechanics free energy maps and nonadiabatic simulations for a photochemical reaction in DNA: cyclobutane thymine dimer. J Phys Chem Lett 7(21):4391–4397

64. Richter M, Marquetand P, González-Vázquez J, Sola I, González L (2011) SHARC: ab initio molecular dynamics with surface hopping in the adiabatic representation including arbitrary couplings. J Chem Theory Comput 7(5):1253–1258
65. Franco de Carvalho F, Tavernelli I (2015) Nonadiabatic dynamics with intersystem crossings: a time-dependent density functional theory implementation. J Chem Phys 143(22):224105
66. Penfold TJ, Curchod BFE, Tavernelli I, Abela R, Rothlisberger U, Chergui M (2012) Simulations of x-ray absorption spectra: the effect of the solvent. Phys Chem Chem Phys 14:9444
67. Tapavicza E, Tavernelli I, Rothlisberger U (2007) Trajectory surface hopping within linear response time-dependent density-functional theory. Phys Rev Lett 98(2):1–4

Chapter 2
The Diplatinum Complex PtPOP

In the present project we have investigated the photocatalytic diplatinum(II) complex $[Pt_2(P_2O_5H_2)_4]^{4-}$, abbreviated PtPOP. The study employed both ultrafast XDS measurements in aqueous solution, performed together with the group of our experimental collaborators, and a combination of gas-phase DFT and QM/MM BOMD simulations. The use of experimental and computational methods has proved highly synergetic: the simulations guided the analysis and interpretation of the XDS data, while the experiments have been a testing ground for fully assessing the potentialities of the ΔSCF-QM/MM BOMD method that has been implemented in the course of the project. Furthermore, the simulations are used to expand on the knowledge of the solution properties of the system and move forward in the understanding of the microscopic mechanisms governing ultrafast relaxation in solution following photoexcitation. In this chapter, we present the model photocatalyst PtPOP, describing the photophysical, structural and dynamical properties that are already known from previous studies, and highlighting the pending questions that we aimed to address in our investigation.

Figure 2.1 shows an illustration of the structure of the PtPOP system. Owing to its nuclear and electronic structures, PtPOP is the prototype system of choice for photophysical studies within a family of highly photoreactive d^8–d^8 binuclear complexes [2]. The UV-vis absorption spectrum of PtPOP in crystal and different solvents exhibits an intense band around 370 nm and a weaker band around 450 nm that are attributed to electronic transition from the highest occupied molecular orbital (HOMO) $d\sigma^*$ antibonding to the lowest unoccupied molecular orbital (LUMO) $p\sigma$ bonding metal-metal orbital [3–5]. As a result of the nature of the excitation, the first singlet and triplet excited states (S_1 and T_1), having $d\sigma^* \rightarrow p\sigma$ character, feature a significantly shortened Pt−Pt distance. Reported experimental values for the contraction in crystal and solution lie in the range 0.19–0.31 Å [4, 6–10]. From the vibronic progression of low temperature UV-vis $S_0 \rightarrow S_1$ and $S_0 \rightarrow T_1$ absorption bands [3], it has been concluded that the potential energy surfaces of S_1 and T_1 along the Pt−Pt coordinate are parallel. Moreover, these states are found, from experi-

Parts of this chapter have been reproduced with permission from Ref. [1], https://doi.org/10.1021/acs.jpcc.8b00301. Copyright 2018 American Chemical Society.

G. Levi, *Photoinduced Molecular Dynamics in Solution*, Springer Theses,
https://doi.org/10.1007/978-3-030-28611-8_2

13

Fig. 2.1 Visualization of the
PtPOP molecular complex.
This photocatalytic model
system has been object of
extensive experimental and
computational investigation
during the present Ph.D.
project

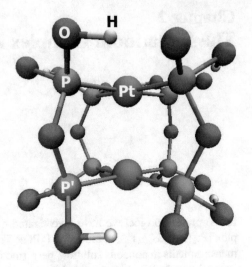

ments [3, 4] and previous DFT studies [11, 12], to be separated by a relatively large
energy gap of around 0.65 eV, and isolated from other electronic states. The electronic structure of the complex, together with the fact that direct spin-orbit coupling
(SOC) between S_1 and T_1 is forbidden for symmetry reasons [11, 13], accounts for
intersystem crossing (ISC) times between 11.0 and 101.5 ps [3, 14, 15], depending
on solvent and temperature. Besides, the lifetime of T_1 is found to be on the order of
microseconds [3]. Ultimately, it is this state that has catalytic activity, being able to
abstract hydrogen and halogen atoms from different substrates [2, 16].

The shape and relative position of the S_1 and T_1 PESs of the complex determine its
unique photophysical properties. Yet, the topology of the PESs has only been deduced
from optical measurements. One of the goals of our investigation was to compute
the PESs along the Pt−Pt coordinate in the first two excited states for the first time.
This represented both a benchmark of the performances of ΔSCF with respect to
structural predictions of transition metal complexes, and an indication that proposed
structures and PES shapes deduced indirectly from optical experiments are indeed
justified. Furthermore, the calculations were also aimed at testing the assumption
made in previous computational works on similar systems [17, 18], which simulated
the singlet excited-state dynamics by using the gradients of the triplet surface.

Previous ultrafast studies have exploited the peculiar photophysical properties
of PtPOP to characterize, by femtosecond optical measurements, the evolution of
coherent wave packet vibrations along the Pt−Pt coordinate in S_1 [14] and recently
also in the T_1 state [19]. Some of the aspects of the ultrafast relaxation following
excitation in the S_1 state in different solvents where uncovered in a combined fluorescence up-conversion and broadband transient absorption study by van der Veen
et al. [14]. It was found that the coherence decay of vibrational wave packets with a
period of ∼224 fs takes place concomitantly with vibrational cooling over a remark-
ably long time of 1–2 ps. The observations were interpreted as a signature of the

strong harmonicity of the potential along the Pt—Pt coordinate, which in turn is due to the rigidity of the cage of P-O-P ligands, and effective shielding from random solvent fluctuations provided by the latter. Despite the fact that the experiments could characterize the time scales of vibrational coherence, cooling and ISC in solution, the mechanistic details behind these processes are far from being well understood. Hypotheses of mechanisms of vibrational cooling have been put forward, but they are not based on direct experimental evidence; rather they rely on the observation of solvent trends [14] or the comparison with the behaviour of diplatinum systems with modified ligands under similar experimental conditions [19]. Thus, van der Veen et al. [14] explain differences in the vibrational decay rates for excitation in the S_1 state observed for different solvents as an evidence of direct solute-solvent interactions that can only occur along the open axial Pt—Pt coordination sites of the molecule. More recently, Monni et al. [19] seem to exclude this possibility. These authors argue that, since no big differences with respect to the decoherence times of a perfluoroborated derivative of PtPOP for which the bulkier ligands offer better shielding of the Pt atoms from the environment were observed, the origin of coherence decay must arise from anharmonic couplings of the Pt—Pt motion with other internal vibrational modes in the complex.

The mechanism of ISC from S_1 to T_1 in PtPOP is also a recurrent subject of discussion in the PtPOP literature [2, 11, 13, 14]. All recent experimental indications seem to point in the direction of a possible involvement of a dark mode that would lower the D_{4h} symmetry of the Pt_2P_8 core of the complex, allowing for direct SOC or lowering the energy of other triplet states, but this mode has never been observed experimentally. The scenario is complicated by the fact that up to now no experimental method has been able to reliably assess the changes affecting the structure of the ligands or the presence of large amplitude distortions in the excited state in solution.

In the light of all this, a second objective of our investigation was to clarify the aspects of the excited-state vibrational relaxation of PtPOP in solution that have remained so far poorly understood, shedding light on questions like: *what is the geometry of the ligand cage in the excited state? Are there ligand deformations that can influence the intersystem crossing rates? What is the role of the solvent in the ultrafast relaxation?* For this purpose, we used ΔSCF in extensive nonequilibrium gas-phase and solution-phase simulations in conjunction with thorough vibrational analysis.

While the excited-state structural dynamics of PtPOP has been object of extensive ultrafast experimental investigations in recent years, no studies exist that address the dynamics in the ground state along the same lines. The present understanding of the ground-state potential surface of the molecule is limited to the observations of the early low-temperature emission [4] and Raman [20] spectroscopic experiments, which deduced a highly harmonic potential along the Pt—Pt coordinate, with a vibrational period of around 303 and 283 fs in crystal [4] and aqueous solution [20], respectively. But, for example, no ultrafast studies have been reported that investigate the vibrational relaxation in the ground state. This is mainly due to the fact that pump-probe techniques are all based on photoexcitation of the sample, and hence usually highlight the excited-state dynamics at the expense of the dynamics occurring in the ground-state molecular ensemble perturbed by the laser. We have participated to an experimental campaign performed at the Linac Coherent Light Source (LCLS) XFEL facility [21, 22] of Stanford to study by time-resolved XDS measurements in water the coherent vibrational dynamics of PtPOP in the ground-state potential. Direct tracking of ground-state dynamics was enabled by a careful choice of pump-pulse parameters to suppress any excited-state contribution in the time dependence of the XDS signal. QM/MM BOMD simulations were subsequently used to substantiate the outcome of the ultrafast XDS experiments.

References

1. Levi G, Pápai M, Henriksen NE, Dohn AO, Møller KB (2018) Solution structure and ultrafast vibrational relaxation of the PtPOP complex revealed by ΔSCF-QM/MM direct dynamics simulations. J. Phys. Chem. C 122:7100–7119
2. Gray HB, Záliš S, Vlček A (2017) Electronic structures and photophysics of d8–d8 complexes. Coord. Chem. Rev. 345:297–317
3. Stiegman AE, Rice SF, Gray HB, Miskowski VM (1987) Electronic spectroscopy of d^8-d^8 diplatinum complexes. $^1a_{2u}(d\sigma^* \to p\sigma)$, $^3e_u(d_{xz},d_{yz} \to p\sigma)$, and $^{3,1}b_{2u}(d\sigma^* \to d_{x^2-y^2})$ excited states of $pt_2(p_2o_5h_2)_4^{4-}$. Inorg Chem 26:1112
4. Rice SF, Gray HB (1983) Electronic absorption and emission spectra of binuclear platinum(II) complexes. Characterization of the lowest singlet and triplet excited states of $Pt_2(P_2O_5H_2)_4^{4-}$. J Am Chem Soc 105:4571–4575
5. Fordyce WA, Brummer JG, Crosby GA (1981) Electronic spectroscopy of a diplatinum(II) octaphosphite complex. J Am Chem Soc 103(6):7061–7064
6. van der Veen RM, Milne CJ, El Nahhas A, Lima FA, Pham VT, Best J, Weinstein JA, Borca CN, Abela R, Bressler C, Chergui M (2009) Structural determination of a photochemically active diplatinum molecule by time-resolved EXAFS spectroscopy. Angew Chem Int Ed 48(15):2711–2714
7. Christensen M, Haldrup K, Bechgaard K, Feidenhans R, Kong Q, Cammarata M, Lo Russo M, Wulff M, Harrit N, Nielsen MM (2008) Time-resolved x-ray scattering of an electronically excited state in solution. Structure of the a state of Tetrakis-μ-pyrophosphitodiplatinate (II). J Am Chem Soc 131(Ii):502–508
8. Yasuda N, Uekusa H, Ohashi Y (2004) X-ray analysis of excited-state structures of the diplatinum complex anions in five crystals with different cations. Bull Chem Soc Jpn 77(5):933–944
9. Kim CD, Pillet S, Wu G, Fullagar WK, Coppens P (2002) Excited-state structure by time-resolved X-ray diffraction. Acta Crystallogr Sect A Found Crystallogr 58:133–137
10. Ikeyama T, Yamamoto S, Azumi T (1988) Vibrational analysis of sublevel phosphorescence spectra of potassium tetrakis(μ-diphosphonato)diplatinate(II): mechanism of radiative transition for the electronically forbidden A1u spectrum. J Phys Chem 92(24):6899
11. Záliš S, Lam Y-C, Gray HB, Vlček A (2015) Spin-orbit TDDFT electronic structure of diplatinum(II, II) complexes. Inorg Chem 54:3491–3500
12. Novozhilova IV, Volkov AV, Coppens P (2003) Theoretical analysis of the triplet excited state of the [Pt2(H2P2O5)4]4- ion and comparison with time-resolved X-ray and spectroscopic results. J Am Chem Soc 125(4):1079–1087
13. Durrell AC, Keller GE, Lam YC, Sýkora J, Vlček A, Gray HB (2012) Structural control of 1A2u-to-3A2u intersystem crossing in diplatinum(II, II) complexes. J Am Chem Soc 134(34):14201–14207

14. van der Veen RM, Cannizzo A, van Mourik F, Vlček Jr A, Chergui M (2011) Vibrational relaxation and intersystem crossing of binuclear metal complexes in solution. J Am Chem Soc 113:305
15. Milder SJ, Brunschwig BS (1992) Factors affecting nonradiative decay: temperature dependence of the picosecond fluorescence lifetime of Pt2(pop)44-. J Phys Chem 96(5):2189–2196
16. Roundhill DM, Gray HB, Che C-M (1989) Pyrophosphito-bridged diplatinum chemistry. Acc Chem Res 22(9):55–61
17. van Driel TB, Kjær KS, Hartsock R, Dohn AO, Harlang T, Chollet M, Christensen M, Gawelda W, Henriksen NE, Kim JG, Haldrup K, Kim KH, Ihee H, Kim J, Lemke H, Sun Z, Sundstrom V, Zhang W, Zhu D, Møller KB, Nielsen MM, Gaffney KJ (2016) Atomistic characterization of the active-site solvation dynamics of a photocatalyst. Nat Commun 7:13678
18. Dohn AO, Jónsson EÖ, Kjær KS, van Driel TB, Nielsen MM, Jacobsen KW, Henriksen NE, Møller KB (2014) Direct dynamics studies of a binuclear metal complex in solution: the interplay between vibrational relaxation, coherence, and solvent effects. J Phys Chem Lett 5:2414–2418
19. Monni R, Auböck G, Kinschel D, Aziz-Lange KM, Gray HB, Vlček A, Chergui M (2017) Conservation of vibrational coherence in ultrafast electronic relaxation: the case of diplatinum complexes in solution. Chem Phys Lett 683:112–120
20. Che CM, Butler LG, Gray HB, Crooks RM, Woodruff WH (1983) Metal-metal interactions in binuclear platinum(II) diphosphite complexes. Resonance Raman spectra of the $1A_{1g}(d\sigma^*)^2$ and $3A_{2u}(d\sigma^*p\sigma)$ electronic states of $(Pt_2(P_2O_5H_2)_4^{4-})$. J Am Chem Soc 105(16):5492–5494
21. Chollet M, Alonso-Mori R, Cammarata M, Damiani D, Defever J, Delor JT, Feng Y, Glownia JM, Langton JB, Nelson S, Ramsey K, Robert A, Sikorski M, Song Sa, Stefanescu D, Srinivasan V, Zhu D, Lemke HT, Fritz DM (2015) The X-ray Pump-Probe instrument at the Linac coherent light source. J Synchrotron Radiat 22:503–507
22. Emma P, Akre R, Arthur J, Bionta R, Bostedt C, Bozek J, Brachmann A, Bucksbaum P, Coffee R, Decker FJ, Ding Y, Dowell D, Edstrom S, Fisher A, Frisch J, Gilevich S, Hastings J, Hays G, Ph Hering Z, Huang R, Iverson H, Loos M, Messerschmidt A, Miahnahri S, Moeller HD, Nuhn G, Pile D, Ratner J, Rzepiela D, Schultz T, Smith P, Stefan H, Tompkins J, Turner J, Welch J, White W, Wu J, Yocky G, Galayda J (2010) First lasing and operation of an ångstrom-wavelength free-electron laser. Nat Photon 4(9):641–647

Chapter 3
Outline of the Thesis

To help the reader find his/her way through the thesis, we summarize here in short, compact form the contents and scopes of each of the following chapters.

Chapter 4 introduces the reader to the theory of nuclear dynamics from the full time-dependent Schrödinger equation to the approximations that form the basis of ab initio on-the-fly Born-Oppenheimer Molecular Dynamics (BOMD).

Chapter 5 delves into the details of the implementation of a ΔSCF method with Gaussian smeared constraints in the DFT code GPAW, realized during the present Ph.D. project. In order to bring out its salient features and differences with respect to other, more standard ΔSCF DFT methods, an effort is made to show the origin of the equations that form its basis, providing all necessary theoretical background on DFT and GPAW. Finally, the chapter reports the results of tests of the newly implemented ΔSCF scheme on a diatomic system that are performed to assess the robustness and reliability of the method with respect to structural predictions.

Chapter 6 describes the QM/MM electrostatic embedding scheme in GPAW/ASE. In addition, it establishes the link between all the components of the QM/MM BOMD simulations performed in the present work.

Chapters 7 and 8 deal with the experimental side of the present project. Chapter 7 provides a broad outline of the principles of time-resolved X-ray diffuse scattering (XDS) experiments, and describes the XDS measurements performed on PtPOP in water at the LCLS XFEL of Stanford. Chapter 8 bridges experiments and theory by showing how the scattering signal can be simulated in order to assist the analysis and interpretation of the experimental data.

Chapter 9 reports the results of preliminary tests and gas-phase calculations on PtPOP using GPAW. This chapter includes the first calculated potential energy surfaces (PESs) of the complex, and highlights the finding that the lowest-lying singlet and triplet excited states of the molecule have a different symmetry than that of the ground state, in contrast to what so far believed.

Chapter 10 sets the stage for the presentation of the results of the QM/MM BOMD simulations on PtPOP in water by illustrating the computational procedure used to

© Springer Nature Switzerland AG 2019
G. Levi, *Photoinduced Molecular Dynamics in Solution*, Springer Theses,
https://doi.org/10.1007/978-3-030-28611-8_3

perform them, focusing in particular on the choice of initial conditions to simulate laser-induced out-of-equilibrium dynamics in the ground and first singlet excited states.

Chapter 11 deals with the equilibrium thermal properties of PtPOP as obtained from the NVT equilibrated parts of the QM/MM trajectories. A detailed characterization of the solvation shell structure is presented, underlining the link with previous experimental evidence. Moreover, this chapter shows how the simulations are used to improve on the structural modelling of the XDS data of PtPOP leading to the first experimental determinantion of the change in Pt–Pt distance in the lowest-lying singlet excited state in water.

Chapter 12 presents a picture of simultaneous ground- and excited-state dynamics following laser excitation obtained through nonequilibrium ΔSCF-QM/MM BOMD simulations and non-stationary ground-state distributions from an equilibrium QM/MM ensemble. The picture shows how the formation of a non-stationary hole in the ground-state distribution of Pt–Pt distances accompanied by a vibrationally "cold" excited state can explain the origin of the oscillatory trend observed in the time-resolved XDS signal.

Chapter 13 presents the results of nonequilibrium ΔSCF-QM/MM BOMD simulations that shed light on the mechanisms of vibrational relaxation of PtPOP in the first singlet excited state in water. This chapter uncovers the paths of dissipation of excess Pt–Pt vibrational energy to ligand deformation modes, and the role of the solvent in stabilizing them.

Part II
Theoretical and Computational Methods

Chapter 4
Nuclear Dynamics

In all formulas and derivations presented in this part of the thesis we will make use of atomic units, in which the electron mass m_e, the elementary charge e and the reduced planck constant $\hbar = h/2\pi$ are unity.

In general, the exact evolution of a non-relativistic molecular system is given by the time dependence of the total electronic and nuclear wave function $|\Psi\rangle$, obtained by solving the time-dependent Schrödinger equation:

$$i\frac{\partial}{\partial t}|\Psi\rangle = \mathbf{H}|\Psi\rangle \tag{4.1}$$

where $\mathbf{H}$ is the total Hamiltonian for coupled electronic-nuclear motion, consisting of a sum of the nuclear kinetic energy operator $\mathbf{T}_n$ and the electronic Hamiltonian $\mathbf{H}_e$. For a system of N_n nuclei and N_e electrons, $\mathbf{H}_e$ can be expressed as:

$$
\begin{aligned}
\mathbf{H}_e &= -\sum_{i=1}^{N_e}\frac{1}{2}\nabla_i^2 - \sum_{i=1}^{N_e}\sum_{\alpha=1}^{N_n}\frac{\mathcal{Z}_\alpha}{|\mathbf{R}_\alpha - \mathbf{r}_i|} + \sum_{i=1}^{N_e}\sum_{j>i}^{N_e}\frac{1}{|\mathbf{r}_i - \mathbf{r}_j|} + \sum_{\alpha=1}^{N_n}\sum_{\beta>\alpha}^{N_n}\frac{\mathcal{Z}_\alpha \mathcal{Z}_\beta}{|\mathbf{R}_\alpha - \mathbf{R}_\beta|}\\
&= \mathbf{T}_e + \mathbf{V}_{ne} + \mathbf{V}_{ee} + \mathbf{V}_{nn}
\end{aligned}
\tag{4.2}
$$

where $\mathbf{R}_\alpha$ and $\mathbf{r}_i$ are respectively the position vectors of nucleus α and electron i, while $\mathcal{Z}_\alpha$ is the charge of nucleus α (corresponding to its atomic number). In Eq. (4.2), the first term is the kinetic energy of the electrons, the second term represents the Coulomb attraction between electrons and nuclei, and the third and fourth terms are the electron-electron and internuclear repulsion, respectively, the latter being a constant for a given nuclear configuration.

Directly finding analytical solutions to Eq. (4.1) is impracticable even for the smallest polyatomic systems. The route to the solution of the problem of deter-

© Springer Nature Switzerland AG 2019
G. Levi, *Photoinduced Molecular Dynamics in Solution*, Springer Theses,
https://doi.org/10.1007/978-3-030-28611-8_4

mining ab initio the dynamics of a molecular system starts from a separation of the electronic and nuclear motions. In fact, owing to the large difference in mass between electrons and nuclei, the time scales of electronic motion are much shorter than those that characterize the motion of the nuclei. Therefore, we can define an electronic Hamiltonian for each set of nuclear positions $\mathbf{R}_\alpha$. The solutions of the time-independent electronic Schrödinger equation for fixed nuclear configurations:

$$\mathbf{H}_e |\Phi_n; \mathbf{R}\rangle = E_n(\mathbf{R}) |\Phi_n; \mathbf{R}\rangle \tag{4.3}$$

are stationary electronic wave functions $|\Phi_n; \mathbf{R}\rangle$ with corresponding energies $E_n(\mathbf{R})$, both dependent parametrically on the collective set of nuclear coordinates $\mathbf{R}$. The total wave function $|\Psi\rangle$ can be exactly expanded in the complete set of these electronic states. In the coordinate representation:

$$\Psi(\mathbf{R}, \mathbf{r}) = \langle \mathbf{R}, \mathbf{r} | \Psi \rangle = \sum_n^\infty \langle \mathbf{R}, \Phi_n; \mathbf{R} | \Psi \rangle \langle \mathbf{r} | \Phi_n; \mathbf{R} \rangle$$

$$= \sum_n^\infty \chi_n(\mathbf{R}) \, \Phi_n(\mathbf{r}; \mathbf{R}) \tag{4.4}$$

Equation (4.4) is the Born-Huang, or adiabatic expansion [1], and defines the $\mathbf{R}$-dependent expansion coefficients $\chi_n(\mathbf{R})$ of the total wavefunction as projections onto a direct product of an eigenstate of the position operator with a particular electronic state $|\mathbf{R}\rangle \otimes |\Phi_\mathbf{n}; \mathbf{R}\rangle = |\mathbf{R}, \Phi_\mathbf{n}; \mathbf{R}\rangle$.

Now, it is understood that the problem of describing the time evolution of a molecular system has been reduced to the determination of the time dependence of the functions $\chi_n(\mathbf{R})$. Obtaining the coefficients $\chi_n(\mathbf{R})$ can be done by solving the following set of coupled differential equations (see Ref. [2] for a complete derivation of this result):

$$i \frac{\partial}{\partial t} \chi_n(\mathbf{R}) = \left[\mathbf{T}_n + E_n(\mathbf{R}) \right] \chi_n(\mathbf{R})$$

$$- \sum_m^\infty \sum_{\alpha=1}^{N_n} \frac{1}{2M_\alpha} \big[\langle \Phi_n; \mathbf{R} | \nabla_\alpha^2 | \Phi_m; \mathbf{R} \rangle$$

$$+ 2 \langle \Phi_n; \mathbf{R} | \nabla_\alpha | \Phi_m; \mathbf{R} \rangle \cdot \nabla_\alpha \big] \chi_m(\mathbf{R}) \tag{4.5}$$

where M_α is the mass of nucleus α. The terms with $n \neq m$ appearing in the double summation over electronic states and nuclei on the right hand side of Eq. (4.5) couple different electronic states through the nuclear motion and define the non-adiabatic quantum dynamics of the system. Terms with $n = m$ are usually called diagonal couplings [2], even if, strictly speaking, they do not couple different electronic states.

4.1 The Born-Oppenheimer Approximation

A considerable simplification of the equations of nuclear motion (4.5) can be achieved by neglecting all non-adiabatic and diagonal coupling terms, obtaining:

$$i\frac{\partial}{\partial t}\chi_n(\mathbf{R}) = \left[\mathbf{T}_n + E_n(\mathbf{R})\right]\chi_n(\mathbf{R}) \tag{4.6}$$

The approximation that we have just introduced is the Born-Oppenheimer (BO) approximation [1]. It implies complete separation of the equations for nuclear and electronic motion. By neglecting all couplings between electronic states it is assumed that the electronic character of the system does not change during nuclear motion, as there cannot be transitions between electronic states. As a consequence, only one term n appears in the expansion of the total wave function Eq. (4.4).

From Eq. (4.6) we can define a Hamiltonian for the motion of the nuclei as the sum of the nuclear kinetic energy operator and the electronic state energy E_n. Thus, in the BO approximation, the eigenvalues E_n of the time-independent electronic Schrödinger equation represent the potential energy surfaces (PESs) on which the nuclei move. One commonly refers to the coefficients $\chi_n(\mathbf{R})$ as nuclear wave functions, although they are not necessarily eigenstates of this nuclear Hamiltonian, rather they can be any superposition of stationary nuclear states satisfying Eq. (4.6) [3].

The BO approximation is widely employed in simulations of molecular systems in which the motion of the nuclei is confined in well separated electronic potentials, far from regions of the electronic and nuclear configuration space where non-adiabatic effects are important.

4.2 Ab Initio Born-Oppenheimer Molecular Dynamics

A further approximation that can be made on the basis of the large mass of the nuclei as compared to that of the electrons, is to describe the dynamics of the nuclei using classical equations of motion. In its most generic formulation, ab initio Born-Oppenheimer Molecular Dynamics (BOMD) [4, 5] propagates a system of atoms in a given adiabatic electronic state n by integrating Newton's equations of motion:

$$\frac{\partial^2 \mathbf{R}_\alpha}{\partial t^2} - \frac{\mathbf{F}_\alpha}{M_\alpha} = 0 \tag{4.7}$$

with forces $\mathbf{F}_\alpha$ computed as the gradients of the eigenvalues of the electronic Schrödinger equation (Eq. (4.3)) for state n ($\mathbf{F}_\alpha = -\nabla_\alpha E_n(\mathbf{R})$). Equations (4.3) and (4.7) are the basic equations of ab initio BOMD simulations.

One strategy involves solving Eq. (4.3) for different nuclear configurations, and fitting the resulting points to an appropriate function to obtain a "global" PES for the classical trajectory propagation. However, obtaining accurate "global" PESs for systems with more than three or four atoms can be extremely challenging [5], thus posing a limitation to the utilization of this method for simulations of the dynamics of large molecular systems. A second strategy consists in solving simultaneously Eqs. (4.3) and (4.7), which means computing, at each step of the classical propagation, ab initio energy and gradients. Ab initio BOMD simulations based on this approach are usually referred to as direct or on-the-fly methods, and allow, in principle, to explore the full, unconstrained space of nuclear configurations.

The present work is concerned with this second strategy as a route to simulate the ground- and excited-state dynamics of systems as large as transition metal complexes, including explicit solvent effects. In Chap. 6 we will see how solvent effects can be taken into account in a multiscale fashion within the scheme presented herein, and how the classical equations of motion can be integrated to reproduce an NVT, or canonical, ensemble. In the following chapter, initially, we will have a closer look at electronic structure methods for solving the time-independent electronic Schrödinger equation based on density functional theory (DFT), with particular focus on the projector augmented wave (PAW) method.

The starting point of all electronic structure methods is the variational principle, which states that the expectation value of the electronic Hamiltonian given any approximate wave function $|\mathbf{\Phi}; \mathbf{R}\rangle$:

$$\langle E \rangle = \frac{\langle \mathbf{\Phi}; \mathbf{R} | \mathbf{H_e} | \mathbf{\Phi}; \mathbf{R} \rangle}{\langle \mathbf{\Phi}; \mathbf{R} | \mathbf{\Phi}; \mathbf{R} \rangle} \tag{4.8}$$

is an upper bound to the exact energy. The variational principle has a disarmingly simple form. For the ground state:

$$\langle E \rangle \geqslant E_0 \tag{4.9}$$

where the equality holds only when $|\mathbf{\Phi}; \mathbf{R}\rangle$ is equal to the exact wave function $|\mathbf{\Phi_0}; \mathbf{R}\rangle$ of the ground state. Hence, obtaining the solutions of the time-independent electronic Schrödinger equation can be done by minimizing the energy as a functional of the electronic wave function, or, as we will see soon, as a functional of the electron density, subject to specific constraints, the nature of which is determined by the choice of the variational parameters.

References

1. Born M, Huang K (1968) Dynamical theory of crystal lattices. Oxford University Press
2. Malhado JP, Bearpark MJ, Hynes JT (2014) Non-adiabatic dynamics close to conical intersections and the surface hopping perspective. Frontiers Chem 2(97):1–21

3. Henriksen NE, Hansen FY (2008) Theories of molecular reaction dynamics. Oxford University Press
4. Marx D, Hutter J (2009) Ab initio molecular dynamics: basic theory and advanced methods. Cambridge University Press
5. Jensen F (2017) Introduction to computational chemistry, 3rd edn. Wiley

Chapter 5
Density Functional Methods

In this chapter, we embark on an excursion into the realm of density functional methods for solving the time-independent electronic Schrödinger equation (Eq. (4.3)). The first part (Sects. 5.1–5.3) is intended to be a general outline of the principles and foundations of density functional theory (DFT), although an effort is made to illustrate the genesis of its workhorse equations. For comprehensive reviews on DFT and step-by-step derivations the reader can consult Refs. [1–3]. The second part (Sects. 5.4 and 5.5) is dedicated to the specific DFT code used during this project, and to the development works done in it.

By examining the expression for the electronic Hamiltonian $\mathbf{H}_e$ given in Eq. (4.2), it is easy to see that for a system of N_e electrons, $\mathbf{H}_e$ is completely specified by the external potential of the nuclei ("external" from the point of view of the electrons):

$$v(\mathbf{r}) = -\sum_{\alpha=1}^{N_n} \frac{\mathcal{Z}_\alpha}{\mid \mathbf{R}_\alpha - \mathbf{r} \mid} \tag{5.1}$$

Therefore, the nuclear charges and positions, which determine $v(\mathbf{r})$, uniquely define the electronic energy and all other properties of a system of N_e electrons. The premises of DFT stem from the simple realization that the electron density, which is a physical observable, provides all the quantities required to construct $v(\mathbf{r})$ and fix the electronic Hamiltonian. Recalling a result of wave mechanics [4], the electron density is obtained from the wave function squared integrated over the $N_e - 1$ electronic spatial coordinates $\mathbf{r}_i$ and the N_e spin coordinates ξ_i. For the ground state:

$$n(\mathbf{r}) = N_e \int \cdots \int \mid \Phi_0\left(\mathbf{r}, \xi_1, \mathbf{x}_2, \ldots, \mathbf{x}_{N_e}; \mathbf{R}\right) \mid^2 d\xi_1 d\mathbf{x}_2 \cdots d\mathbf{x}_{N_e} \tag{5.2}$$

where we have introduced the notation $\mathbf{x}_i$ to indicate the collection of spatial and spin coordinates for electron i ($\mathbf{x}_i = \{\mathbf{r}_i, \xi_i\}$). We see that $n(\mathbf{r})$ is a function of three

© Springer Nature Switzerland AG 2019
G. Levi, *Photoinduced Molecular Dynamics in Solution*, Springer Theses,
https://doi.org/10.1007/978-3-030-28611-8_5

variables that integrates to the total number of electrons:

$$\int n(\mathbf{r})d\mathbf{r} = N_e \tag{5.3}$$

Moreover, the positions and charges of the nuclei can be inferred [3], respectively, from the positions of local cusps in the density and from the relation:

$$\left.\frac{\partial}{\partial d_\alpha}\bar{n}(d_\alpha)\right|_{d_\alpha \to 0^+} = -2\mathcal{Z}_\alpha \bar{n}(0) \tag{5.4}$$

where d_α is the radial distance from nucleus α and $\bar{n}$ is the density averaged over a sphere.

5.1   The Hohenberg-Kohn Theorems

In this and the following two sections we will lay out the standard DFT formalism for the electronic ground state, leaving the discussion of excited states to Sect. 5.5.

Since all information that is needed to determine the electronic Hamiltonian can be deduced from the electron density, there must be a one-to-one correspondence between $n(\mathbf{r})$ and the electronic energy corresponding to the exact wave function. The formal justification that the electron density can be used as basic variable in solving the electronic Schrödinger equation is provided by the two Hohenberg-Kohn theorems [5].

The first theorem is a proof that the external potential $v(\mathbf{r})$, and hence the electronic wave function and energy of the ground state, are uniquely determined by the electron density. The demonstration is done by reductio ad absurdum using the variational principle for the ground state (see Eqs. (4.8) and (4.9)). For ease of notation we will drop from now on the parametric dependence of the electronic wave function on the collective set of nuclear coordinates $\mathbf{R}$. Let us assume there exist two external potentials $v^{(a)}(\mathbf{r})$ and $v^{(b)}(\mathbf{r})$ associated with the same ground-state electron density $n(\mathbf{r})$. The two potentials are not necessarily Coulomb potentials set by the nuclei, but have to be one-electron operators. $v^{(a)}(\mathbf{r})$ and $v^{(b)}(\mathbf{r})$ define two different electronic Hamiltonians $\mathbf{H}_e^{(a)}$ and $\mathbf{H}_e^{(b)}$, and two different ground-state wave functions $|\Phi_0^{(a)}\rangle$ and $|\Phi_0^{(b)}\rangle$, which are taken to be normalized ($\langle\Phi_0^{(a)}|\Phi_0^{(a)}\rangle = \langle\Phi_0^{(b)}|\Phi_0^{(b)}\rangle = 1$). The variational principle for $|\Phi_0^{(b)}\rangle$ with respect to the Hamiltonian $\mathbf{H}_e^{(a)}$ gives:

$$\langle\Phi_0^{(b)}|\mathbf{H}_e^{(a)}|\Phi_0^{(b)}\rangle > E_0^{(a)} \tag{5.5}$$

By rewriting the left hand side of Eq. (5.5) as:

$$\langle \Phi_0^{(b)} | \mathbf{H}_e^{(a)} | \Phi_0^{(b)} \rangle = \langle \Phi_0^{(b)} | \mathbf{H}_e^{(b)} | \Phi_0^{(b)} \rangle + \langle \Phi_0^{(b)} | \mathbf{H}_e^{(a)} - \mathbf{H}_e^{(b)} | \Phi_0^{(b)} \rangle$$

$$= E_0^{(b)} + \int n(\mathbf{r}) \left[v^{(a)}(\mathbf{r}) - v^{(b)}(\mathbf{r}) \right] d\mathbf{r} \tag{5.6}$$

where the second equality comes from the fact that $v^{(a)}(\mathbf{r})$ and $v^{(b)}(\mathbf{r})$ are one-electron operators, we arrive at the following expression:

$$E_0^{(b)} + \int n(\mathbf{r}) \left[v^{(a)}(\mathbf{r}) - v^{(b)}(\mathbf{r}) \right] d\mathbf{r} > E_0^{(a)} \tag{5.7}$$

Analogously, we could repeat the derivation using $|\Phi_0^{(a)}\rangle$ as an approximate wave function for $\mathbf{H}_e^{(b)}$, obtaining:

$$E_0^{(a)} - \int n(\mathbf{r}) \left[v^{(a)}(\mathbf{r}) - v^{(b)}(\mathbf{r}) \right] d\mathbf{r} > E_0^{(b)} \tag{5.8}$$

Adding Eqs. (5.7) and (5.7) on both sides gives:

$$E_0^{(b)} + E_0^{(a)} > E_0^{(a)} + E_0^{(b)} \tag{5.9}$$

which is obviously an impossible conclusion, showing that the density $n(\mathbf{r})$ must define a single external potential, and hence a unique Hamiltonian and a unique ground-state wave function $|\Phi_0\rangle$ with associated energy E_0. The important implication of this result is that we can express the electronic energy of the system as a unique functional of the density $(E[n])$.

By analogy with the definition of the electronic Hamiltonian in Eq. (4.2), we can separate the total energy functional in the following terms:

$$E[n] = T_e[n] + V_{ne}[n] + V_{ee}[n] + V_{nn} \tag{5.10}$$

where $T_e[n]$ is the electronic kinetic energy, $V_{ne}[n]$ and $V_{ee}[n]$ are, respectively, the Coulomb attraction between electrons and nuclei and the electron-electron interaction:

$$V_{ne}[n] = -\sum_{\alpha=1}^{N_n} \int \frac{Z_\alpha n(\mathbf{r})}{|\mathbf{R}_\alpha - \mathbf{r}|} d\mathbf{r} = \int v(\mathbf{r}) n(\mathbf{r}) d\mathbf{r} \tag{5.11}$$

$$V_{ee}[n] = J[n] + \text{xc term}$$

$$= \frac{1}{2} \int \int \frac{n(\mathbf{r}) n(\mathbf{r}')}{|\mathbf{r} - \mathbf{r}'|} d\mathbf{r} d\mathbf{r}' + \text{xc term} \tag{5.12}$$

and, finally, V_{nn} is the constant (within the BO approximation) internuclear repulsion. In Eq. (5.12) we have separated the classical electron-electron repulsion $(J[n])$ from a nonclassical term, which makes up the major portion of the so-called exchange-correlation (xc) energy of the interacting system of electrons.

The second Hohenberg-Kohn theorem gives the perscription for how to evaluate the energy of the ground state from the electronic Schrödinger equation using the electron density. It is basically an energy variational principle for the electron density. Consider an approximate electron density $n'(\mathbf{r})$ that is positive definite ($n'(\mathbf{r}) \geqslant 0$) and integrates to the total number of electrons ($\int n'(\mathbf{r})d\mathbf{r} = N_e$). Then, the value of the energy functional of this approximate density will be greater than or equal to the true ground-state energy:

$$E\left[n'\right] \geqslant E_0\left[n\right] \tag{5.13}$$

The idea of using a function of only three variables as variational parameter to minimize the energy of a molecular system is particularly appealing in view of a reduction of the complexity brought about by the $4N_e$ variables ($3N_e$ spatial and N_e spin coordinates) of the electronic wave function in wave mechanics.

Moving along these lines, we can reformulate the electronic Schrödinger equation (Eq. (4.3)) as the problem of minimizing the functional $E[n]$ with respect to the electron density:

$$\delta E\left[n\right] = \int \frac{\delta E\left[n\right]}{\delta n(\mathbf{r})} \delta n(\mathbf{r})d\mathbf{r} = 0 \tag{5.14}$$

or equivalently:

$$\frac{\delta E\left[n\right]}{\delta n(\mathbf{r})} = 0 \tag{5.15}$$

where, in Eq. (5.14), we have used the definition of differential of a functional, and $\dfrac{\delta E\left[n\right]}{\delta n(\mathbf{r})}$ is the functional derivative of $E[n]$ with respect to $n(\mathbf{r})$. Minimization should be carried out under the constraint that $n(\mathbf{r})$ integrates to the total number of electrons:

$$\int n(\mathbf{r})d\mathbf{r} - N_e = 0. \tag{5.16}$$

This problem can be solved using the method of Lagrange multipliers [3], which leads to the following equations:

$$\frac{\delta}{\delta n(\mathbf{r})}\left[E\left[n\right] - \mu\left(\int n(\mathbf{r})d\mathbf{r} - N_e\right)\right] = 0$$

$$\Rightarrow \frac{\delta E\left[n\right]}{\delta n(\mathbf{r})} - \mu = 0 \tag{5.17}$$

Equation (5.17) is called Euler-Lagrange equation and the Lagrange multiplier μ is the chemical potential. By inserting the definition of the total energy functional Eq. (5.10) into Eq. (5.17), and using the expression for the classical attraction between the electron density and the nuclei given in Eq. (5.11), we finally obtain:

$$\frac{\delta}{\delta n(\mathbf{r})} \left[T_e[n] + V_{ee}[n] \right] + v(\mathbf{r}) = \frac{\delta F[n]}{\delta n(\mathbf{r})} + v(\mathbf{r}) = \mu \qquad (5.18)$$

where we have defined the energy functional $F[n]$ as the sum of the electronic kinetic energy functional $T_e[n]$ and the electron-electron interaction term $V_{ee}[n]$. $F[n]$ is a *universal functional* of the electron density, in that it does not dependent on the external potential $v(\mathbf{r})$. If we knew the form of $F[n]$ we could exactly solve Eq. (5.18) for the electron density and, thus, determine the true electronic energy of a system of atoms by inserting the resulting $n(\mathbf{r})$ into Eq. (5.10). Unfortunately, the functional $F[n]$ is not known, and DFT does not provide any indication on how we might proceed to find the exact form of it.

The lack of knowledge of $F[n]$ poses severe limitations to the applicability of orbital-free DFT to molecular and solid systems of interest. Historically, there have been attempts to develop orbital-free density functional models for a uniform electron gas (the so-called Thomas-Fermi and Thomas-Fermi-Dirac theories [1–3]), however these models fail to predict bondings between atoms. Efforts to try to overcome this challenge that are based on finding strategies to construct density functionals using machine learning [6] are currently being undertaken. The idea is pursued, in particular, by the group of Burke [6, 7]. Such machine learning density functional methods have only very recently started to move their first steps from one-dimensional systems to simulations of realistic molecular systems [8].

5.2 The Kohn-Sham Equations

At this point, we could have turned our backs on DFT if it were not for Kohn and Sham, who, in 1965, presented a formulation of DFT, the Kohn-Sham (KS) DFT method [9], that has found, and continues to find, wide spread use in many different sectors of science [10].

The method brings into play a wave function expressed as a single Slater determinant:

$$\Phi(\mathbf{x}_1, \mathbf{x}_2, \ldots, \mathbf{x}_{N_e}) = \frac{1}{\sqrt{N_e!}} \begin{vmatrix} \psi_1(\mathbf{x}_1) & \psi_2(\mathbf{x}_1) & \cdots & \psi_{N_e}(\mathbf{x}_1) \\ \psi_1(\mathbf{x}_2) & \psi_2(\mathbf{x}_2) & \cdots & \psi_{N_e}(\mathbf{x}_2) \\ \vdots & \vdots & \ddots & \vdots \\ \psi_1(\mathbf{x}_{N_e}) & \psi_2(\mathbf{x}_{N_e}) & \cdots & \psi_{N_e}(\mathbf{x}_{N_e}) \end{vmatrix}$$

$$= |\psi_1(\mathbf{x}_1)\psi_2(\mathbf{x}_2)\cdots\psi_{N_e}(\mathbf{x}_{N_e})\rangle$$

$$= |\psi_1\psi_2\cdots\psi_{N_e}\rangle \qquad (5.19)$$

where each $\psi_i(\mathbf{x})$ is a spin orbital given by the product of a spatial orbital $\phi_i(\mathbf{r})$ and a spin function $\alpha(\xi)$ or $\beta(\xi)$. The spatial orbitals and the spin functions are assumed to be orthonormal. This single-determinant wave function is the exact wave function for the ground state of a system of N_e *noninteracting* electrons. The exact kinetic energy of a system of noninteracting electrons can be expressed using a set of N_{sorb} spin orbitals (with $N_{sorb} \geqslant N_e$) as:

$$T_e^s[n] = \sum_{i=1}^{N_{sorb}} f_i \langle \psi_i | -\frac{1}{2}\nabla_i^2 | \psi_i \rangle \tag{5.20}$$

The system also has an exact electron density that is given by [3, 4]:

$$n(\mathbf{r}) = \sum_{i=1}^{N_{sorb}} f_i \, | \, \psi_i(\mathbf{x}) \, |^2 \tag{5.21}$$

In Eqs. (5.20) and (5.21), the f_i are occupation numbers for the orbitals. For the ground state, the assignment of the occupation numbers follows the aufbau principle, i.e. f_i is equal to 1 for the lowest energy orbitals, and 0 for all other orbitals. Note that in Eq. (5.20) we have indicated the kinetic energy as an implicit functional of the electron density through Eq. (5.21).

The main idea underlying KS DFT is to use Eqs. (5.20) and (5.21) to express the kinetic energy and density of a real system of *interacting* electrons. The resulting KS total energy functional for the real system takes the following form:

$$E_{KS}[n] = T_e^s[n] + V_{ne}[n] + J[n] + E_{xc}[n] + V_{nn} \tag{5.22}$$

where the exchange-correlation energy functional $E_{xc}[n]$ is defined as the difference between the exact kinetic energy of the interacting system of electrons ($T_e[n]$) and $T_e^s[n]$, plus all nonclassical contributions to the electron-electron interaction energy ($V_{ee}[n] - J[n]$, as seen from Eq. (5.12)):

$$E_{xc}[n] = T_e[n] - T_e^s[n] + V_{ee}[n] - J[n] \tag{5.23}$$

Thus, the exchange-correlation energy consists of a correction to account for the interacting nature of the true system. The exact form of $E_{xc}[n]$ is not known. However, the correction is, in most cases, small compared to the absolute value of the kinetic energy $T_e^s[n]$ [1, 3], such that an approximate $E_{xc}[n]$ usually suffices to achieve fairly accurate results in many cases.

By inserting the definitions of the terms $T_e^s[n]$, $V_{ne}[n]$ and $J[n]$ given by Eqs. (5.20), (5.11) and (5.12), respectively, we can rewrite Eq. (5.23) as:

$$E_{KS}[n] = \sum_{i=1}^{N_{sorb}} f_i \langle \psi_i | -\frac{1}{2}\nabla_i^2 |\psi_i\rangle + \int v(\mathbf{r})n(\mathbf{r})d\mathbf{r} + \frac{1}{2}\int\int \frac{n(\mathbf{r})n(\mathbf{r}')}{|\mathbf{r}-\mathbf{r}'|}d\mathbf{r}d\mathbf{r}'$$
$$+ E_{xc}[n] + V_{nn} \tag{5.24}$$

Since the density is obtained from the orbitals $\psi_i(\mathbf{x})$, which from now on will be referred to as "KS orbitals", the Hohenberg-Kohn variational problem (Eqs. (5.14)–(5.17)) becomes, in the framework of the KS theory, the problem of minimizing the value of the energy functional $E_{KS}[n]$ with respect to the $\psi_i(\mathbf{x})$, under the constraint that they are orthonormal:

$$\int \psi_i^*(\mathbf{x})\psi_j(\mathbf{x})d\mathbf{x} = \delta_{ij} \tag{5.25}$$

where δ_{ij} is the Kronecker delta. This leads to the following equations:

$$\frac{\delta}{\delta\psi_i^*(\mathbf{x})}\left[E_{KS}[n] - \sum_{i=1}^{N_{sorb}}\sum_{j=1}^{N_{sorb}}\epsilon_{ij}'\left(\int\psi_i^*(\mathbf{x})\psi_j(\mathbf{x})d\mathbf{x} - \delta_{ij}\right)\right] = 0 \tag{5.26}$$

where $\psi_i^*(\mathbf{x})$ is the complex conjugate of $\psi_i(\mathbf{x})$, and the ϵ_{ij}' are Lagrange multipliers. With the explicit definition of $E_{KS}[n]$, Eq. (5.24), inside Eq. (5.26), after computing the functional derivatives with respect to the $\psi_i^*(\mathbf{x})$, we obtain a set of nonlinear coupled equations:

$$f_i \mathbf{h}_{KS}\psi_i(\mathbf{x}) = \sum_{j=1}^{N_{sorb}}\epsilon_{ij}'\psi_j(\mathbf{x}) \tag{5.27}$$

where the single-particle KS Hamiltonian $\mathbf{h}_{KS}$ is defined as:

$$\mathbf{h}_{KS} = -\frac{1}{2}\nabla_i^2 + v_{KS}(\mathbf{r}) \tag{5.28}$$

with the effective KS potential $v_{KS}(\mathbf{r})$ given as a sum of three terms:

$$v_{KS}(\mathbf{r}) = v(\mathbf{r}) + v_H(\mathbf{r}) + v_{xc}(\mathbf{r}) \tag{5.29}$$

The first term is the usual external potential of the nuclei ($v(\mathbf{r})$), $v_H(\mathbf{r})$ is the so-called Hartree potential:

$$v_H(\mathbf{r}) = \int \frac{n(\mathbf{r}')}{|\mathbf{r}-\mathbf{r}'|}d\mathbf{r} \tag{5.30}$$

and the exchange-correlation potential $v_{xc}(\mathbf{r})$ is the functional derivative of the exchange-correlation energy with respect to the electron density:

$$v_{xc}(\mathbf{r}) = \frac{\delta E_{xc}[n]}{\delta n(\mathbf{r})} \qquad (5.31)$$

Since the matrix ϵ' with elements ϵ'_{ij} is a Hermitian matrix [3], we can apply a unitary transformation of the KS orbitals that diagonalizes ϵ' while leaving invariant the wave function $\Phi(\mathbf{x})$ of Eq. (5.19) and the Hamiltonian $\mathbf{h}_{KS}$ [3, 4]. If we do so, we obtain from (5.27) a new set of simplified equations:

$$\mathbf{h}_{KS}\psi_i(\mathbf{x}) = \epsilon_i \psi_i(\mathbf{x}) \qquad (5.32)$$

where $\epsilon_i = \epsilon'_{ii}/f_i$ for $f_i \neq 0$. These equations are termed KS equations and must be solved for the KS orbitals iteratively, until convergence of the electron density, since the density appears in the expression of the Hamiltonian $\mathbf{h}_{KS}$ (Eq. (5.28)). Thus, just like the Hartree-Fock (HF) theory of wave mechanics [4], KS DFT relies on a self-consistent field (SCF) procedure to obtain the orbitals that minimize the total energy. Once these are available, the energy can be determined by first computing the electron density according to Eq. (5.21) and then inserting the result into Eq. (5.24).

As for how to determine the KS orbitals in practice, different strategies are available. The route most commonly followed by quantum chemists (which is also the method of choice in wave function theories as HF) is to expand the spatial part of the KS orbitals in a basis of localized functions $\zeta_\mu(\mathbf{r})$ resembling atomic orbitals:

$$|\phi_i\rangle = \sum_\mu c_{i\mu}|\zeta_\mu\rangle \qquad (5.33)$$

The basis set functions are usually taken as linear combinations of Gaussian-type orbitals (GTOs). The coefficients appearing in the expansion of the KS orbitals are determined by solving the matrix equation obtained from the variational procedure in the atomic orbital basis set. The computational cost, in this case, scales as N_b^4, with N_b the number of basis functions employed in the calculation.

However, this is not the only method for solving the KS equations. An expansion of the orbitals in a plane-wave basis set is also possible. This is the approach that is usually preferred in the solid-state physics community to model periodic systems [1]. Since plane waves are unsuited to describe the strong localization and rapidly varying nodal structure of the core orbitals, Effective Core Potentials (ECPs), or Pseudo-Potentials (PPs), are needed to model the core electrons [1]. Sometimes, as is the case for part of the calculations that will be presented in this work, ECPs are also used in conjunction with an explicit description of the valence electrons in terms of localized orbital basis sets. Considerable savings of computational time can be achieved when employing ECPs for heavy atoms, such as transition metal complexes, instead of localized basis set functions, because, otherwise, the number of basis functions that would be required is very large [1]. Moreover, relativistic effects can be taken into account by fitting the analytical form of the ECPs to results from reference relativistic calculations.

A drawback of PP methods is that all information on the electronic structure of a system near the nuclei are lost. The Projector Augmented Wave (PAW) method [11, 12] is a third alternative strategy to the solution of the KS equations that allows, in principle, to retain all core properties at a computational cost that is comparable to the one offered by PP approaches. The theory of the PAW method can be derived as an exact theory, and is treated extensively in the following section (Sect. 5.4).

Another aspect that one must consider before venturing into the "black box" of KS DFT calculations, is the choice of xc functional. As already mentioned, the form of this functional has to be approximated. The literature offers an overwhelming amount of different xc functionals. Some of them, like PBE [13], are the result of a rational design following a set of conditions that a functional is required to satisfy. Most often, however, the functionals are constructed by fitting some parameters to accurate experimental data. The very popular BLYP [14, 15] functional, for example, belongs to this other class of xc functionals.

As there is no unique parameter that can be varied to systematically increase the accuracy of the xc functionals, a classification of them is not easy. On the other hand, *it is* possible to define a hierarchy of density functional approximations [1, 16] on the basis of the "ingredients" used in the preparation of the xc functionals. The simple rule is: adding more "ingredients" is expected to give increasingly improved functionals. At the bottom of the ladder of density functional approximations we find the local density approximation (LDA), which makes the xc functional depend exclusively on the local values of the electron density. The LDA exchange is the exchange energy of a uniform electron gas, for which an exact analytical form exists. The most common LDA correlation functionals can be traced back to the VWN [17] and PW [18] parametrizations, which have been fitted to accurate quantum Monte Carlo results. The next level of approximation is to make the functional depend also on the gradients of the density (generalized-gradient approximation (GGA)). Popular GGA exchange functionals are B86 [19], B88 [14] and PBE (exchange) [13]; while among the GGA correlation functionals we can mention PW91 [20], PBE (correlation) [13] and LYP [15]. The name of an xc functional is, usually, and in particular for GGA functionals, constructed by merging the acronyms for the exchange and correlation parts; so, for example, BLYP is B88 exchange plus LYP correlation. The direct QM/MM simulations of PtPOP in water performed in the present work made use of the GGA functional BLYP to describe the electronic structure of the complex. Finally, at the high rungs of the ladder we find hybrid functionals, such as B3LYP [21, 22], that include some portion of exact HF exchange energy. The results of B3LYP calculations on PtPOP are used, in the present work, to asses the quality of the geometry of the complex as predicted by BLYP.

5.3   Restricted and Unrestricted Formalisms

Here, we provide some definitions concerning the construction of Slater determinants to represent the electronic wave function in KS DFT, most of them valid also in HF theory. We introduce some concepts and notations that will be used throughout this thesis, especially when discussing DFT calculations for excited states (Sect. 5.5).

One can conceive different kinds of Slater determinants, depending on the type of constraint that is enforced on the spatial part of the spin orbitals $\psi_i(\mathbf{x})$.

In the restricted formalism, spin orbitals with α and β spin functions are constrained to have the same spatial part. Let us consider first a system with an even number of electrons N_e, and where all spatial orbitals are doubly occupied, meaning that for each of them there will be two electrons. The set of N_e spin orbitals that form the determinant is obtained from $N_e/2$ spatial orbitals by multiplying each of them once by a spin function α ($\psi_{i-1}(\mathbf{x}) = \phi_{i/2}(\mathbf{r})\alpha(\xi)$, $i = 2, 4, \ldots, N_e$), and once by a spin function β ($\psi_i(\mathbf{x}) = \phi_{i/2}(\mathbf{r})\beta(\xi)$, $i = 2, 4, \ldots, N_e$). The determinant thus obtained is a restricted closed-shell determinant (RCS). An example of such determinant is given in Fig. 5.1 for a four electron system. Using the short-hand notation introduced in Eq. (5.19), we can write a general restricted closed-shell determinant as:

$$^1\mathbf{\Phi}_{RCS}\left(\mathbf{x}_1, \mathbf{x}_2, \ldots, \mathbf{x}_{N_e-1}, \mathbf{x}_{N_e}\right) = |\psi_1\psi_2 \cdots \psi_{N_e-1}\psi_{N_e}\rangle$$
$$= |\phi_1\bar{\phi}_1 \cdots \phi_{N_e/2}\bar{\phi}_{N_e/2}\rangle \qquad (5.34)$$

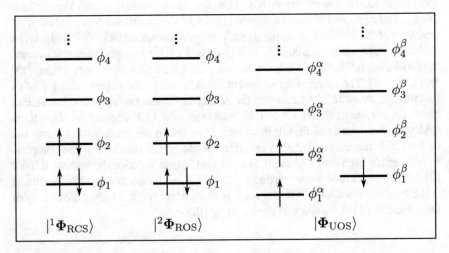

Fig. 5.1 Examples of single Slater determinants. From left to right: restricted closed-shell (RCS), doublet restricted open-shell (ROS), and approximate doublet unrestricted open-shell (UOS) determinants

where, in the last line, we have further introduced a notation in which spin orbitals are indicated with their spatial part only ($\bar{\phi}_i$ corresponds to a spin orbital containing a β spin function). The superscript on the left of Φ_{RCS} in Eq. (5.34) tells us that the determinant is a singlet, i.e. its spin multiplicity given by $2S + 1$, where S is the spin angular momentum quantum number, is 1. This means that Φ_{RCS} is an eigenfunction of the square of the total spin angular momentum operator $\mathbf{S}$:

$$\mathbf{S}^2|\Phi_{RCS}\rangle = S(S+1)|\Phi_{RCS}\rangle = 0 \tag{5.35}$$

Furthermore, as *any* single determinant [4], Φ_{RCS} is an eigenfunction of the z component of the total spin operator:

$$\mathbf{S}_z|\Phi_{RCS}\rangle = M_S|\Phi_{RCS}\rangle = 0 \tag{5.36}$$

(in general, for a single determinant $M_S = \dfrac{N_e^\alpha - N_e^\beta}{2}$, where N_e^α and N_e^β are, respectively, the number of α and β electrons). The electron density of a Slater determinant of the form of Φ_{RCS} is given, after integrating out the spin functions (compare with Eq. (5.21)), by:

$$n(\mathbf{r}) = 2 \sum_{i=1}^{N_e/2} |\phi_i(\mathbf{r})|^2 \tag{5.37}$$

where the occupation number 2 in front of the summation derives from the fact that each spatial orbital $\phi_i(\mathbf{r})$ is doubly occupied.

Next we consider a system with an odd number of electrons. The determinant describing this system is necessarily an open-shell determinant, since there is at least one unpaired electron. We might use the restricted formalism also in this case. Figure 5.1 shows a restricted open-shell (ROS) determinant of three electrons with one unpaired electron. A restricted open-shell determinant with one unpaired electron is a doublet, and can be written as:

$$^2\Phi_{ROS}(\mathbf{x}_1, \mathbf{x}_2, \ldots, \mathbf{x}_{N_e-1}, \mathbf{x}_{N_e}) = |\phi_1\bar{\phi}_1 \cdots \bar{\phi}_{(N_e-1)/2}\phi_{(N_e+1)/2}\rangle \tag{5.38}$$

Note, however, that not all open-shell restricted determinants are eigenfunctions of $\mathbf{S}^2$. We will see examples of such cases in Sect. 5.5 when treating determinants with open-shell electrons of different spin.

In the open-shell determinant given by Eq. (5.38), electrons with spin α experience a different exchange potential than the β electrons, due to $N_e^\alpha \neq N_e^\beta$ and the fact that exchange interactions are only between electrons with the same spin. Therefore, we might expect the spatial part of the α spin orbitals to be different from that of the β spin orbitals. In the unrestricted formalism, α and β spin orbitals are allowed to have different spatial parts. Thus, the spin orbitals are constructed from a set of $\phi_i^\alpha(\mathbf{r})$ and a set of $\phi_i^\beta(\mathbf{r})$ spatial orbitals (see the example in Fig. 5.1). The unrestricted open-shell

(UOS) determinant for the case where only one electron is unpaired can be written as:

$$\Phi_{\text{UOS}}\left(\mathbf{x}_1, \mathbf{x}_2, \ldots, \mathbf{x}_{N_e-1}, \mathbf{x}_{N_e}\right) = |\phi_1^\alpha \overline{\phi}_1^\beta \cdots \overline{\phi}_{(N_e-1)/2}^\beta \phi_{(N_e+1)/2}^\alpha\rangle \tag{5.39}$$

The lack of superscript on the left of Φ_{UOS} hints at the fact that this determinant is not an eigenfunction of $\mathbf{S}^2$ (even though it is an eigenfunction of $\mathbf{S}_z$ with $M_S = \frac{1}{2}$). This can be generalized to any unrestricted determinant [4]. As a consequence, unrestricted determinants are not pure spin states, but contain contaminations of higher spin multiplicities. Nevertheless, unrestricted determinants are usually taken as first approximations to pure spin states. In a KS DFT calculation employing the unrestricted formalism, two different sets of KS equations need to be solved, one for the $\phi_i^\alpha(\mathbf{r})$ and one for the $\phi_i^\beta(\mathbf{r})$ spatial orbitals.

One can always define an electron density for α electrons and an electron density for β electrons that summed give the total density. For an unrestricted determinant:

$$n(\mathbf{r}) = n^\alpha(\mathbf{r}) + n^\beta(\mathbf{r}) = \sum_{i=1}^{N_e^\alpha} |\phi_i^\alpha(\mathbf{r})|^2 + \sum_{i=1}^{N_e^\beta} |\phi_i^\beta(\mathbf{r})|^2 \tag{5.40}$$

Usually, one defines also a spin density $n_s(\mathbf{r})$ as given by the difference between the α and β densities:

$$n_s(\mathbf{r}) = n^\alpha(\mathbf{r}) - n^\beta(\mathbf{r}) = \sum_{i=1}^{N_e^\alpha} |\phi_i^\alpha(\mathbf{r})|^2 - \sum_{i=1}^{N_e^\beta} |\phi_i^\beta(\mathbf{r})|^2 \tag{5.41}$$

Exchange-correlation functionals can be formulated in terms of $n^\alpha(\mathbf{r})$ and $n^\beta(\mathbf{r})$. For open-shell systems, often (but not always, as we will see in Sect. 5.5), DFT calculations employ spin-polarized functionals. The expressions "unrestricted" and "spin-polarized" are usually used interchangeably to indicate DFT calculations with unrestricted determinants.

5.4   The Projector Augmented Wave Method

The PAW method has already been briefly mentioned in Sect. 5.2, where it has been presented as a strategy to solve the KS DFT equations with a computational cost similar to that of PP methods, but that, contrary to the latter, formally preserves all aspects of the wave function, and electron density, in the core regions. One of the difficulties connected with electronic structure calculations, in general, is to account for the rapid oscillations exhibited by the orbitals near the nuclei. We will see how the PAW approach bypasses this problem by introducing smooth auxiliary orbitals as variational parameters in the SCF minimization procedure; and by doing that in a way that allows to reconstruct the full KS orbitals with the correct nodal structure near the nuclei. Before delving into the practical aspects of PAW calculations, we shall shortly review the formalism underlying the method starting from its basic principles. For more exhaustive descriptions of the methodology, Refs. [11, 12] are ideal starting points.

5.4.1 Pseudo Orbitals

In what follows, the PAW theory is presented using exclusively spatial orbitals, as the spin part of the KS spin orbitals are, in practice, not amenable to numerical computation and can be integrated out at any time [4].

We seek a linear transformation $\mathcal{T}$ that can map the full KS orbitals $\phi_i(\mathbf{r})$ into smooth auxiliary orbitals $\tilde{\phi}_i(\mathbf{r})$:

$$|\phi_i\rangle = \mathcal{T}|\tilde{\phi}_i\rangle \tag{5.42}$$

The papers where the method was first presented (Refs. [11, 12]), used the terminology "wave function" to indicate both the $|\phi_i\rangle$ and the $|\tilde{\phi}_i\rangle$ one-particle functions. This nomenclature has also been by the GPAW program [23, 24]. Here, to be consistent with the terminology used in Sect. 5.2, and avoid confusion with the many-particle electronic wave function, we will continue to call them "orbitals". We also drop the term "all-electron" to indicate the KS orbitals $|\phi_i\rangle$, which in the referenced articles is used to distinguish them from the $|\tilde{\phi}_i\rangle$. However, the auxiliary orbitals $|\tilde{\phi}_i\rangle$, and all other quantities directly connected to them, will be given the attribute of "pseudo", as in the original formulation.

We require that the operator $\mathcal{T}$ transforms the pseudo orbitals only within augmentation spheres surrounding the nuclei, such that we can write it as the identity plus some local atom-centered operators $\mathcal{T}^\alpha$:

$$\mathcal{T} = 1 + \sum_{\alpha=1}^{N_n} \mathcal{T}^\alpha \tag{5.43}$$

This form of the transformation operator implies that the pseudo orbitals are equal to the KS orbitals outside the augmentation spheres. The equivalence is justified by the fact that between atoms, in the bonding regions, the KS orbitals are already smooth, and, therefore, there is no need to apply the transformation there.

Then, we expand the KS orbitals inside each augmentation region α in terms of a complete basis set of partial waves $|\varphi_\mu^\alpha\rangle$:

$$|\phi_i\rangle = \sum_\mu^\infty c_{i\mu}^\alpha |\varphi_\mu^\alpha\rangle, \quad \text{within } |\mathbf{r} - \mathbf{R}_\alpha| < r_c^\alpha \tag{5.44}$$

where r_c^α is the cutoff radius defining the augmentation region of atom α, and the expansion coefficients $c_{i\mu}^\alpha$ are to be determined. Next, we associate to each partial wave a smooth counterpart $|\tilde{\varphi}_\mu^\alpha\rangle$, termed pseudo partial wave. The $|\tilde{\varphi}_\mu^\alpha\rangle$ for atom α are chosen such that they are related to the partial waves $|\varphi_\mu^\alpha\rangle$ through the transformation $\mathcal{T}^\alpha$:

$$|\varphi_\mu^\alpha\rangle = (1 + \mathcal{T}^\alpha) |\tilde{\varphi}_\mu^\alpha\rangle$$
$$\Rightarrow \mathcal{T}^\alpha |\tilde{\varphi}_\mu^\alpha\rangle = |\varphi_\mu^\alpha\rangle - |\tilde{\varphi}_\mu^\alpha\rangle \tag{5.45}$$

Since $\mathcal{T}^\alpha$ operates inside the augmentation regions only, we have that:

$$|\varphi_\mu^\alpha\rangle = |\tilde{\varphi}_\mu^\alpha\rangle, \quad \text{for } |\,\mathbf{r} - \mathbf{R}_\alpha\,| > r_c^\alpha \tag{5.46}$$

We, then, take the $|\tilde{\varphi}_\mu^\alpha\rangle$ as basis in which to expand the pseudo orbitals inside the augmentation regions:

$$|\tilde{\phi}_i\rangle = \sum_\mu^\infty c_{i\mu}^\alpha |\tilde{\varphi}_\mu^\alpha\rangle, \quad \text{within } |\,\mathbf{r} - \mathbf{R}_\alpha\,| < r_c^\alpha \tag{5.47}$$

Since:

$$|\varphi_\mu^\alpha\rangle = \mathcal{T} |\tilde{\varphi}_\mu^\alpha\rangle \tag{5.48}$$

we see, by inserting the above expression for $|\varphi_\mu^\alpha\rangle$ into Eq. (5.44) and comparing the result with Eq. (5.42), that the coefficients of the expansions in Eqs. (5.47) and (5.44) must be identical. In order to make the transformation operator $\mathcal{T}$ linear, these expansion coefficients are taken as the scalar products of the pseudo orbitals $|\tilde{\phi}_i\rangle$ with some localized functions $|\tilde{p}_\mu^\alpha\rangle$, called projector functions:

$$c_{i\mu}^\alpha = \langle \tilde{p}_\mu^\alpha | \tilde{\phi}_i \rangle \tag{5.49}$$

By inserting Eq. (5.49) into Eq. (5.47), we obtain the following expression for the pseudo orbitals:

$$|\tilde{\phi}_i\rangle = \sum_\mu^\infty |\tilde{\varphi}_\mu^\alpha\rangle \langle \tilde{p}_\mu^\alpha | \tilde{\phi}_i \rangle, \quad \text{within } |\,\mathbf{r} - \mathbf{R}_\alpha\,| < r_c^\alpha \tag{5.50}$$

which implies the identity relation:

$$\sum_\mu^\infty |\tilde{\varphi}_\mu^\alpha\rangle \langle \tilde{p}_\mu^\alpha | = 1, \quad \text{within } |\,\mathbf{r} - \mathbf{R}_\alpha\,| < r_c^\alpha \tag{5.51}$$

and that:

$$\langle \tilde{p}^\alpha_\mu | \tilde{\varphi}^\alpha_\nu \rangle = \delta_{\mu\nu}, \quad \text{within } | \mathbf{r} - \mathbf{R}_\alpha | < r^\alpha_c \tag{5.52}$$

To derive an expression for $\mathcal{T}$, we first operate with $\mathcal{T}^\alpha$ on $|\tilde{\phi}_i\rangle$:

$$\mathcal{T}^\alpha |\tilde{\phi}_i\rangle = \sum_\mu^\infty \mathcal{T}^\alpha |\tilde{\varphi}^\alpha_\mu\rangle \langle \tilde{p}^\alpha_\mu | \tilde{\phi}_i\rangle = \sum_\mu^\infty \left(|\varphi^\alpha_\mu\rangle - |\tilde{\varphi}^\alpha_\mu\rangle \right) \langle \tilde{p}^\alpha_\mu | \tilde{\phi}_i\rangle \tag{5.53}$$

where the first equality comes from using Eq. (5.50), and the second equality from the second line of Eq. (5.45). Equation (5.53) gives the following definition of $\mathcal{T}^\alpha$:

$$\mathcal{T}^\alpha = \sum_\mu^\infty \left(|\varphi^\alpha_\mu\rangle - |\tilde{\varphi}^\alpha_\mu\rangle \right) \langle \tilde{p}^\alpha_\mu | \tag{5.54}$$

Finally, $\mathcal{T}$ is obtained by inserting Eq. (5.54) into Eq. (5.42):

$$\mathcal{T} = 1 + \sum_{\alpha=1}^{N_n} \sum_\mu^\infty \left(|\varphi^\alpha_\mu\rangle - |\tilde{\varphi}^\alpha_\mu\rangle \right) \langle \tilde{p}^\alpha_\mu | \tag{5.55}$$

With the definition of $\mathcal{T}$, we can express the KS orbitals in terms of pseudo orbitals and the partial wave expansions. From Eqs. (5.42) and (5.55):

$$|\phi_i\rangle = |\tilde{\phi}_i\rangle + \sum_{\alpha=1}^{N_n} \sum_\mu^\infty \left(|\varphi^\alpha_\mu\rangle - |\tilde{\varphi}^\alpha_\mu\rangle \right) \langle \tilde{p}^\alpha_\mu | \tilde{\phi}_i\rangle$$

$$= |\tilde{\phi}_i\rangle + \sum_{\alpha=1}^{N_n} \left(|\phi^\alpha_i\rangle - |\tilde{\phi}^\alpha_i\rangle \right) \tag{5.56}$$

where we have defined $|\phi^\alpha_i\rangle$ and $|\tilde{\phi}^\alpha_i\rangle$ as the atom-centered expansions:

$$|\phi^\alpha_i\rangle = \sum_\mu^\infty |\varphi^\alpha_\mu\rangle \langle \tilde{p}^\alpha_\mu | \tilde{\phi}_i\rangle \tag{5.57}$$

$$|\tilde{\phi}^\alpha_i\rangle = \sum_\mu^\infty |\tilde{\varphi}^\alpha_\mu\rangle \langle \tilde{p}^\alpha_\mu | \tilde{\phi}_i\rangle \tag{5.58}$$

From an examination of Eq. (5.56) it should be clear that (i) outside the augmentation regions, due to $|\varphi^\alpha_\mu\rangle = |\tilde{\varphi}^\alpha_\mu\rangle$, the original KS orbitals are equal to the pseudo orbitals

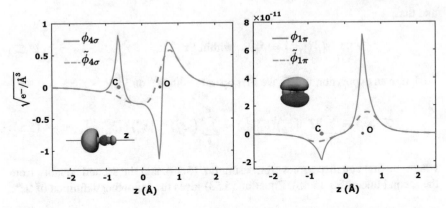

Fig. 5.2 Examples of pseudo (green dashed lines) and KS orbitals (red lines) of a CO molecule obtained from a single-point PAW calculation of the gas-phas ground-state molecule using GPAW with an LDA functional. The plots show the values of the orbitals along the **z** axis, which coincides with the axis of the molecule. Note how, in the outer regions and in between the two atoms, the KS orbitals match their pseudo orbital counterparts; while, close to the nuclei, the pseudo orbitals replace the cusps and oscillating features with smooth continuations

$(|\phi_i\rangle = |\tilde{\phi}_i\rangle)$, and (ii) within the augmentation regions the original KS orbitals are equal to the expansions $|\phi_i^\alpha\rangle$ (Eq. (5.57)), because the pseudo orbitals are equal to the expansions $|\tilde{\phi}_i^\alpha\rangle$ (Eq. (5.58)) as a consequence of Eq. (5.50).

The important achievement that we have attained is a mapping of the full KS problem into one where the variational parameters can be smooth auxiliary functions (the pseudo orbitals $\tilde{\phi}_i(\mathbf{r})$), which are computationally convenient to handle: convergence of a plane-wave or localized atomic orbital basis set is fast for systems with many electrons, and, when using grid-based techniques, as in GPAW (see paragraph below), they can be efficiently represented on coarse grids. On the other hand, the oscillatory nodal structure near the nuclei can be exactly recovered, as Eq. (5.56) suggests. Naturally, the mapping implies that all energy functionals of Eq. (5.22), and the KS Hamiltonian $\mathbf{h}_{KS}$ (Eqs. (5.28)–(5.31)) need to be transformed accordingly for the KS procedure to lead to the correct solution. This will be the topic of the next paragraphs. Examples of pseudo orbitals are given in Fig. 5.2 for a σ and a π orbitals of carbon monoxide (CO), as computed from the isolated ground state of the molecule using GPAW with an LDA functional.

For the PAW method to be exact the basis sets of partial waves $|\varphi_\mu^\alpha\rangle$ and $|\tilde{\varphi}_\mu^\alpha\rangle$, and projectors $|\tilde{p}_\mu^\alpha\rangle$, need to be complete. For practical calculations, however, one truncates the expansions. Usually, one or two partial waves and corresponding projectors per atomic site and angular momentum are sufficient to achieve convergence [12, 23]. The partial waves are obtained from solving the KS equations for the isolated spherically symmetric atoms, often taking into account scalar-relativistic effects [12, 24]. More information on the construction of partial waves and projector functions can be found in Refs. [12, 23]. A second approximation that is usually introduced in practical PAW calculations, is the frozen core approximation. That is to say, only partial waves of valence electrons are included in the expansions of the pseudo and KS orbitals (Eqs. (5.44) and (5.47)), while the orbitals for the core states are fixed to the core partial waves of the isolated atoms ($|\varphi_\mu^{\alpha,\text{core}}\rangle$ and $|\tilde{\varphi}_\mu^{\alpha,\text{core}}\rangle$).

5.4.2 PAW Formulation of the Electron Density

Here, we confine ourselves to reporting the expressions used to compute the electron density within the PAW method. The formulas take into account the approximations that we have introduced above, namely the finite truncation of the partial wave expansions and the frozen-core approximation. Contributions from the core electrons will be separated out to make this clear. For the full derivations see Refs. [11, 12].

The electron density is given in terms of a pseudo density $\tilde{n}(\mathbf{r})$, which is smooth everywhere in space, and atom-centered expansions. For N_n nuclei and N_{orb} spatial orbitals with occupation numbers f_i:

$$n(\mathbf{r}) = \sum_{i=1}^{N_{\text{orb}}} f_i \mid \phi_i(\mathbf{r}) \mid^2 = \tilde{n}(\mathbf{r}) + \sum_{\alpha=1}^{N_n} \left(n^\alpha(\mathbf{r}) - \tilde{n}^\alpha(\mathbf{r}) \right) \qquad (5.59)$$

The pseudo density is obtained from the pseudo orbitals describing N_{eval} valence electrons and a smooth pseudo core density $\tilde{n}_{\text{core}}(\mathbf{r})$:

$$\tilde{n}(\mathbf{r}) = \sum_{i=1}^{N_{\text{orb}}} f_i \mid \tilde{\phi}_i(\mathbf{r}) \mid^2 + \tilde{n}_{\text{core}}(\mathbf{r}) \qquad (5.60)$$

The atom-centered corrections $n^\alpha(\mathbf{r})$ and $\tilde{n}^\alpha(\mathbf{r})$ appearing in Eq. (5.59) are evaluated, for each atom α, from the partial waves and projector functions as:

$$n^\alpha(\mathbf{r}) = \sum_{\mu,\nu} D_{\mu\nu}^\alpha \varphi_\mu(\mathbf{r}) \varphi_\nu(\mathbf{r}) + n_{\text{core}}^\alpha(\mathbf{r}) \qquad (5.61)$$

$$\tilde{n}^\alpha(\mathbf{r}) = \sum_{\mu,\nu} D_{\mu\nu}^\alpha \tilde{\varphi}_\mu(\mathbf{r}) \tilde{\varphi}_\nu(\mathbf{r}) + \tilde{n}_{\text{core}}^\alpha(\mathbf{r}) \qquad (5.62)$$

where $D_{\mu\nu}^\alpha$ are elements of an atomic density matrix, defined as:

$$D_{\mu\nu}^\alpha = \sum_{i=1}^{N_{\text{orb}}} f_i c_{i\mu}^{\alpha*} c_{i\nu}^\alpha = \sum_{i=1}^{N_{\text{orb}}} f_i \langle \tilde{\phi}_i | \tilde{p}_\mu^\alpha \rangle \langle \tilde{p}_\nu^\alpha | \tilde{\phi}_i \rangle \qquad (5.63)$$

and $n_{\text{core}}^\alpha(\mathbf{r})$ and $\tilde{n}_{\text{core}}^\alpha(\mathbf{r})$ are the atomic core electron density and its smooth counterpart. The PAW transformation ensures that each term $n^\alpha(\mathbf{r}) - \tilde{n}^\alpha(\mathbf{r})$ in Eq. (5.59) is non vanishing only inside the augmentation spheres around the nuclei, such that outside them $n(\mathbf{r}) = \tilde{n}(\mathbf{r})$.

5.4.3 The PAW Energy Functional

In order to derive the set of equations that in the PAW method replace the standard KS equations (Eq. (5.32)), which is the ultimate goal of this section, we first have to find the expression of the PAW total energy functional.

Just as for the orbitals (Eq. (5.56)) and the electron density (Eq. (5.56)), also the energy functional can be decomposed into a "smooth" part $\tilde{E}$ plus some atomic corrections:

$$E_{PAW} = \tilde{E} + \sum_{\alpha=1}^{N_n} \left(E^\alpha - \tilde{E}^\alpha \right) \tag{5.64}$$

To see how the above decomposition arises, we need to consider the effects of the PAW transformation on each energy functional term appearing in Eq. (5.22). For some of them, like the kinetic energy functional, we will simply give the final expressions. The terms that arise from Coulomb interactions, on the other hand, will be explicitly derived. This gives us the opportunity to introduce concepts that will be also used in deriving the equations that are at the heart of the QM/MM electrostatic embedding scheme presented in Chap. 6.

The following expression is used in the PAW method to evaluate the kinetic energy functional $T_e^s[n]$ [11, 12, 23]:

$$T_e^s[n] = T_e^s[\tilde{n}] + \sum_{\alpha=1}^{N_n} \left(T_e^{s\alpha} - \tilde{T}_e^{s\alpha} \right) \tag{5.65}$$

where each term is given by:

$$T_e^s[\tilde{n}] = \sum_{i=1}^{N_{orb}} f_i \langle \tilde{\phi}_i | -\frac{1}{2}\nabla_i^2 | \tilde{\phi}_i \rangle \tag{5.66}$$

$$T_e^{s\alpha} = \sum_{\mu,\nu} D_{\mu\nu}^\alpha \langle \varphi_\mu^\alpha | -\frac{1}{2}\nabla_i^2 | \varphi_\nu^\alpha \rangle + \sum_{\mu=1}^{N_{core}^\alpha} \langle \varphi_\mu^{\alpha,core} | -\frac{1}{2}\nabla_i^2 | \varphi_\mu^{\alpha,core} \rangle \tag{5.67}$$

$$\tilde{T}_e^{s\alpha} = \sum_{\mu,\nu} D_{\mu\nu}^\alpha \langle \tilde{\varphi}_\mu^\alpha | -\frac{1}{2}\nabla_i^2 | \tilde{\varphi}_\nu^\alpha \rangle \tag{5.68}$$

with N_{core}^α the number of core states included in the atomic reference calculation for atom α.

For LDA and GGA exchange-correlation functionals, the following general expression applies [11, 12, 23]:

$$
\begin{aligned}
E_{xc}[n] &= E_{xc}[\tilde{n}] + \sum_{\alpha=1}^{N_n} \left(E_{xc}[n^{\alpha}] - E_{xc}[\tilde{n}^{\alpha}] \right) \\
&= E_{xc}[\tilde{n}] + \sum_{\alpha=1}^{N_n} \left(E_{xc}^{\alpha} - \tilde{E}_{xc}^{\alpha} \right)
\end{aligned}
\tag{5.69}
$$

Again, the atomic corrections depend on the density matrix elements $D_{\mu\nu}^{\alpha}$ through Eqs. (5.61) and (5.62).

The way Coulomb electrostatic interactions are handled in the PAW approach is worth more careful consideration. First, a total negative nuclear charge density is defined as the sum of point charge densities $Z^{\alpha}(\mathbf{r})$ of the nuclei, each given by a delta function operating at the nuclear site $\mathbf{R}_{\alpha}$ times the (positive) nuclear charge $\mathcal{Z}_{\alpha}$:

$$
Z(\mathbf{r}) = \sum_{\alpha=1}^{N_n} Z^{\alpha}(\mathbf{r}) = -\sum_{\alpha=1}^{N_n} \delta(\mathbf{r} - \mathbf{R}_{\alpha})\mathcal{Z}_{\alpha}
\tag{5.70}
$$

With this definition of the nuclear density, we can express a total charge density (electron density plus nuclear charge density), which we call $\rho(\mathbf{r})$, as:

$$
\rho(\mathbf{r}) = n(\mathbf{r}) + Z(\mathbf{r}) = n(\mathbf{r}) + \sum_{\alpha=1}^{N_n} Z^{\alpha}(\mathbf{r})
\tag{5.71}
$$

Since $n(\mathbf{r})$ is positive while $Z(\mathbf{r})$ has been defined as a negative quantity, $\rho(\mathbf{r})$ is a sign-inverted charge density, which gives 0 when integrated over all space for a neutral system. Furthermore, we can write the total Coulomb energy, comprising the attraction between electrons and nuclei, and the electron-electron and internuclear repulsion, as the following double integral:

$$
E'_{coul}[n] = \frac{1}{2} \int \int \frac{\rho(\mathbf{r})\rho(\mathbf{r}')}{|\mathbf{r} - \mathbf{r}'|} d\mathbf{r}d\mathbf{r}'
\tag{5.72}
$$

where the prime for $E'_{coul}[n]$ indicates that $E'_{coul}[n]$, as expressed above, includes the infinite self interaction energy between nuclear point charges. Obviously, this term needs to be subtracted out. To avoid excessive notation at this stage, however, we shall apply the correction at a later step. That $E'_{coul}[n]$ is the total Coulomb interaction energy plus the self interaction of the nuclei can be seen by inserting the definition of the total charge density (Eq. (5.71)) into Eq. (5.72):

$$
\begin{aligned}
E'_{\text{coul}}[n] &= \frac{1}{2} \int \int \frac{\rho(\mathbf{r})\rho(\mathbf{r}')}{|\mathbf{r}-\mathbf{r}'|} d\mathbf{r} d\mathbf{r}' \\
&= \frac{1}{2} \int \int \frac{\left(n(\mathbf{r}) - \sum_{\alpha=1}^{N_n} \delta(\mathbf{r}-\mathbf{R}_\alpha)\mathcal{Z}_\alpha\right)\left(n(\mathbf{r}') - \sum_{\beta=1}^{N_n} \delta(\mathbf{r}'-\mathbf{R}_\beta)\mathcal{Z}_\beta\right)}{|\mathbf{r}-\mathbf{r}'|} d\mathbf{r} d\mathbf{r}' \\
&= -\sum_{\alpha=1}^{N_n} \int \frac{\mathcal{Z}_\alpha n(\mathbf{r})}{|\mathbf{R}_\alpha-\mathbf{r}|} d\mathbf{r} + \frac{1}{2} \int \int \frac{n(\mathbf{r})n(\mathbf{r}')}{|\mathbf{r}-\mathbf{r}'|} d\mathbf{r} d\mathbf{r}' + \frac{1}{2}\sum_{\alpha=1}^{N_n}\sum_{\beta=1}^{N_n} \frac{\mathcal{Z}_\alpha \mathcal{Z}_\beta}{|\mathbf{R}_\alpha-\mathbf{R}_\beta|} \\
&= V_{\text{ne}}[n] + J[n] + V'_{\text{nn}}
\end{aligned}
\tag{5.73}
$$

The first term on the third line of Eq. (5.73) is the classical attraction between electrons and nuclei, as in Eq. (5.11), the second term is exactly the electron-electron repulsion as defined in Eq. (5.12), and V'_{nn} is the internuclear repulsion including the self-interaction error.

By inserting the PAW formulation of the electron density (Eq. (5.59)) into Eq. (5.72) and grouping terms that involve a summation over nuclei we obtain:

$$
E'_{\text{coul}}[n] = \frac{1}{2}\left(\left(\tilde{n} + \sum_{\alpha=1}^{N_n}\left(n^\alpha + Z^\alpha - \tilde{n}^\alpha\right)\right)\right)
\tag{5.74}
$$

Here and in what follows, we have introduced the following notation for double integrals:

$$
(f|f') = \int \int \frac{f(\mathbf{r})f'(\mathbf{r}')}{|\mathbf{r}-\mathbf{r}'|} d\mathbf{r} d\mathbf{r}'
\tag{5.75}
$$

$$
(f|f) = ((f))
\tag{5.76}
$$

The expression in Eq. (5.74) can be simplified by introducing a new set of smooth atom-centered functions $\tilde{Z}^\alpha(\mathbf{r})$ localized inside the augmentation spheres:

$$
E'_{\text{coul}}[n] = \frac{1}{2}\left(\left(\tilde{n} + \sum_{\alpha=1}^{N_n}\tilde{Z}^\alpha + \sum_{\alpha=1}^{N_n}\left(n^\alpha + Z^\alpha - \tilde{n}^\alpha - \tilde{Z}^\alpha\right)\right)\right)
\tag{5.77}
$$

and requiring that, by construction of the $\tilde{Z}^\alpha(\mathbf{r})$, the densities $n^\alpha(\mathbf{r}) + Z^\alpha(\mathbf{r}) - \tilde{n}^\alpha(\mathbf{r}) - \tilde{Z}^\alpha(\mathbf{r})$, which vanish outside the augmentation regions, have zero electrostatic multipole moments. As a result, none of the augmentation regions interact electrostatically with the others and the total Coulomb interaction reduces to:

$$
E'_{\text{coul}}[n] = \frac{1}{2}\left(\left(\tilde{n} + \sum_{\alpha=1}^{N_n}\tilde{Z}^\alpha\right)\right) + \frac{1}{2}\sum_{\alpha=1}^{N_n}\left[((n^\alpha + Z^\alpha)) - \left(\left(\tilde{n}^\alpha + \tilde{Z}^\alpha\right)\right)\right]
\tag{5.78}
$$

where we begin to recognize the familiar separation into a "smooth" part and atom-centered corrections. The functions $\tilde{Z}^\alpha(\mathbf{r})$ are usually called compensation charges. For Eq. (5.78) to be exact, the compensation charges should be complete expansions in multipole moments. In GPAW, for practical applications, the expansions are truncated up to the quadrupole moment [25]. The reader interested in the more technical details of how exactly the compensation charges are constructed is referred to Refs. [11, 12, 23]. Here, it will be sufficient to say that they are also functions of the atomic density matrix elements (Eq. (5.63)).

At this point, we can easily get rid of the self-interaction of the nuclear point charges by subtracting a term $\frac{1}{2}\sum_{\alpha=1}^{N_n}((Z^\alpha))$:

$$
\begin{aligned}
E_{\text{coul}}[n] &= E'_{\text{coul}}[n] - \frac{1}{2}\sum_{\alpha=1}^{N_n}((Z^\alpha)) \\
&= \frac{1}{2}\left(\left(\tilde{n} + \sum_{\alpha=1}^{N_n}\tilde{Z}^\alpha\right)\right) + \frac{1}{2}\sum_{\alpha=1}^{N_n}\left[((n^\alpha)) + 2(n^\alpha|Z^\alpha) - \left(\left(\tilde{n}^\alpha + \tilde{Z}^\alpha\right)\right)\right] \\
&= \frac{1}{2}((\tilde{\rho})) + \frac{1}{2}\sum_{\alpha=1}^{N_n}\left[((n^\alpha)) + 2(n^\alpha|Z^\alpha) - \left(\left(\tilde{n}^\alpha + \tilde{Z}^\alpha\right)\right)\right] \\
&= E_{\text{coul}}[\tilde{\rho}] + \sum_{\alpha=1}^{N_n}\left(E^\alpha_{\text{coul}} + \tilde{E}^\alpha_{\text{coul}}\right)
\end{aligned}
\tag{5.79}
$$

where, on the last line, we have defined $E_{\text{coul}}[\tilde{\rho}]$ as the (true) Coulomb energy functional of a pseudo total charge density $\tilde{\rho}(\mathbf{r})$ given by $\tilde{\rho}(\mathbf{r}) = \tilde{n}(\mathbf{r}) + \sum_{\alpha=1}^{N_n}\tilde{Z}^\alpha(\mathbf{r})$. The last equality in Eq. (5.79) defines the three basic components of the Coulomb energy functional. Using the standard notation for the double integrals:

$$
E_{\text{coul}}[\tilde{\rho}] = \frac{1}{2}\int\int\frac{\tilde{\rho}(\mathbf{r})\tilde{\rho}(\mathbf{r}')}{|\mathbf{r} - \mathbf{r}'|}d\mathbf{r}d\mathbf{r}'
\tag{5.80}
$$

$$
E^\alpha_{\text{coul}} = \frac{1}{2}\sum_{\alpha=1}^{N_n}\left[\int\int\frac{n^\alpha(\mathbf{r})n^\alpha(\mathbf{r}')}{|\mathbf{r} - \mathbf{r}'|}d\mathbf{r}d\mathbf{r}' + 2\int\int\frac{n^\alpha(\mathbf{r})Z^\alpha(\mathbf{r}')}{|\mathbf{r} - \mathbf{r}'|}d\mathbf{r}d\mathbf{r}'\right]
\tag{5.81}
$$

$$
\tilde{E}^\alpha_{\text{coul}} = \frac{1}{2}\int\int\frac{\left(\tilde{n}^\alpha(\mathbf{r}) + \tilde{Z}^\alpha(\mathbf{r})\right)\left(\tilde{n}^\alpha(\mathbf{r}') + \tilde{Z}^\alpha(\mathbf{r}')\right)}{|\mathbf{r} - \mathbf{r}'|}d\mathbf{r}d\mathbf{r}'
\tag{5.82}
$$

Having rewritten all the terms appearing in Eq. (5.22) using the PAW formalism, we can, finally, collect them to obtain the expression of the PAW total energy functional:

$$E_{\text{PAW}} = T_e^s [\tilde{n}] + E_{\text{coul}} [\tilde{\rho}] + E_{\text{xc}} [\tilde{n}]$$

$$+ \sum_{\alpha=1}^{N_n} \left(T_e^{s\alpha} - \tilde{T}_e^{s\alpha} + E_{\text{coul}}^{\alpha} - \tilde{E}_{\text{coul}}^{\alpha} + E_{\text{xc}}^{\alpha} - \tilde{E}_{\text{xc}}^{\alpha} \right) \qquad (5.83)$$

By comparing Eq. (5.83) to Eq. (5.64), provided at the beginning of this paragraph, we can now see that:

$$\tilde{E} = T_e^s [\tilde{n}] + E_{\text{coul}} [\tilde{\rho}] + E_{\text{xc}} [\tilde{n}] \qquad (5.84)$$

and:

$$E^{\alpha} = T_e^{s\alpha} + E_{\text{coul}}^{\alpha} + E_{\text{xc}}^{\alpha} \qquad (5.85)$$

$$\tilde{E}^{\alpha} = \tilde{T}_e^{s\alpha} + \tilde{E}_{\text{coul}}^{\alpha} + \tilde{E}_{\text{xc}}^{\alpha} \qquad (5.86)$$

5.4.4 The PAW Hamiltonian

With the expression of the total PAW energy functional at hand we can obtained a set of transformed KS equations by invoking the variational principle. We need to minimize E_{PAW} with respect to the pseudo orbitals under the constraint that the KS orbitals are orthonormal (the pseudo orbitals *do not*, actually, need to be orthonormal):

$$\int \phi_i^*(\mathbf{r}) \phi_j(\mathbf{r}) d\mathbf{r} = \int \tilde{\phi}_i^*(\mathbf{r}) \boldsymbol{T}^\dagger \boldsymbol{T} \tilde{\phi}_j(\mathbf{r}) d\mathbf{r} = \int \tilde{\phi}_i^*(\mathbf{r}) \boldsymbol{O} \tilde{\phi}_j(\mathbf{r}) d\mathbf{r} = \delta_{ij} \qquad (5.87)$$

where $\boldsymbol{O} = \boldsymbol{T}^\dagger \boldsymbol{T}$ is an overlap operator. By applying, as usual, the method of Lagrange multipliers:

$$\frac{\delta}{\delta \tilde{\phi}_i^*(\mathbf{r})} \left[E_{\text{PAW}} - \sum_{i=1}^{N_{\text{orb}}} \sum_{j=1}^{N_{\text{orb}}} \epsilon_{ij}' \left(\int \tilde{\phi}_i^*(\mathbf{r}) \boldsymbol{O} \tilde{\phi}_j(\mathbf{r}) d\mathbf{r} - \delta_{ij} \right) \right] = 0 \qquad (5.88)$$

we obtain, after unitary transformation to diagonalize the matrix of Lagrange multipliers, as done also in deriving Eq. (5.32), the following transformed KS equations:

$$\tilde{\mathbf{h}}_{\text{KS}} \tilde{\phi}_i(\mathbf{r}) = \epsilon_i \boldsymbol{O} \tilde{\phi}_i(\mathbf{r}) \qquad (5.89)$$

where $\tilde{\mathbf{h}}_{\text{KS}} = \boldsymbol{T}^\dagger \mathbf{h}_{\text{KS}} \boldsymbol{T}$ is the transformed KS Hamiltonian, whose explicit form can be derived from the relation:

$$f_i \tilde{\mathbf{h}}_{\text{KS}} \tilde{\phi}_i(\mathbf{r}) = \frac{\delta E_{\text{PAW}}}{\delta \tilde{\phi}_i^*(\mathbf{r})} \qquad (5.90)$$

The functional derivative $\dfrac{\delta E_{\mathrm{PAW}}}{\delta \tilde{\phi}_i^*(\mathbf{r})}$ is evaluated using the definition of the total energy functional E_{PAW} contained in Eqs. (5.64) and (5.84), the expressions for its components $T_{\mathrm{e}}^{\mathrm{s}}[\tilde{n}]$ (Eq. (5.66)) and $E_{\mathrm{coul}}[\tilde{\rho}]$ (Eq. (5.80)), and the definition of $D_{\mu\nu}^{\alpha}$ in Eq. (5.63). We should also keep in mind that $E_{\mathrm{coul}}[\tilde{\rho}]$ and $\left[E^{\alpha} - \tilde{E}^{\alpha}\right]$ are functions of the functionals $D_{\mu\nu}^{\alpha}$ ($E_{\mathrm{coul}}[\tilde{\rho}]$ through the compensation charges). Then, by applying the chain rule for functional derivatives and derivatives of functions of functionals[1]:

$$
\begin{aligned}
\frac{\delta E_{\mathrm{PAW}}}{\delta \tilde{\phi}_i^*(\mathbf{r})} &= \frac{\delta \tilde{E}}{\delta \tilde{\phi}_i^*(\mathbf{r})} + \sum_{\alpha=1}^{N_{\mathrm{n}}} \frac{\delta \left(E^{\alpha} - \tilde{E}^{\alpha}\right)}{\delta \tilde{\phi}_i^*(\mathbf{r})} \\
&= \frac{\delta T_{\mathrm{e}}^{\mathrm{s}}[\tilde{n}]}{\delta \tilde{\phi}_i^*(\mathbf{r})} + \int \left[\frac{\delta E_{\mathrm{coul}}[\tilde{\rho}]}{\delta \tilde{n}(\mathbf{r}')} + \frac{\delta E_{\mathrm{xc}}[\tilde{n}]}{\delta \tilde{n}(\mathbf{r}')}\right] \frac{\delta \tilde{n}(\mathbf{r}')}{\delta \tilde{\phi}_i^*(\mathbf{r})} d\mathbf{r}' \\
&\quad + \sum_{\alpha=1}^{N_{\mathrm{n}}} \sum_{\mu,\nu} \left\{\frac{\partial E_{\mathrm{coul}}[\tilde{\rho}]}{\partial D_{\mu\nu}^{\alpha}} + \frac{\partial \left[E^{\alpha} - \tilde{E}^{\alpha}\right]}{\partial D_{\mu\nu}^{\alpha}}\right\} \frac{\delta D_{\mu\nu}^{\alpha}}{\delta \tilde{\phi}_i^*(\mathbf{r})} \\
&= \frac{\delta T_{\mathrm{e}}^{\mathrm{s}}[\tilde{n}]}{\delta \tilde{\phi}_i^*(\mathbf{r})} + \int \left[\int \frac{\delta E_{\mathrm{coul}}[\tilde{\rho}]}{\delta \tilde{\rho}(\mathbf{r}'')} \frac{\delta \rho(\mathbf{r}'')}{\delta \tilde{n}(\mathbf{r}')} d\mathbf{r}'' + \frac{\delta E_{\mathrm{xc}}[\tilde{n}]}{\delta \tilde{n}(\mathbf{r}')}\right] \frac{\delta \tilde{n}(\mathbf{r}')}{\delta \tilde{\phi}_i^*(\mathbf{r})} d\mathbf{r}' \\
&\quad + \sum_{\alpha=1}^{N_{\mathrm{n}}} \sum_{\mu,\nu} \left\{\int \frac{\delta E_{\mathrm{coul}}[\tilde{\rho}]}{\delta \tilde{\rho}(\mathbf{r})} \frac{\partial \tilde{\rho}(\mathbf{r})}{\partial D_{\mu\nu}^{\alpha}} d\mathbf{r} + \frac{\partial \left[E^{\alpha} - \tilde{E}^{\alpha}\right]}{\partial D_{\mu\nu}^{\alpha}}\right\} \frac{\delta D_{\mu\nu}^{\alpha}}{\delta \tilde{\phi}_i^*(\mathbf{r})} \\
&= \frac{\delta T_{\mathrm{e}}^{\mathrm{s}}[\tilde{n}]}{\delta \tilde{\phi}_i^*(\mathbf{r})} + \left[\frac{\delta E_{\mathrm{coul}}[\tilde{\rho}]}{\delta \tilde{\rho}(\mathbf{r})} + \frac{\delta E_{\mathrm{xc}}[\tilde{n}]}{\delta \tilde{n}(\mathbf{r})}\right] \tilde{\phi}_i(\mathbf{r}) \\
&\quad + \sum_{\alpha=1}^{N_{\mathrm{n}}} \sum_{\mu,\nu} \left\{\int \frac{\delta E_{\mathrm{coul}}[\tilde{\rho}]}{\delta \tilde{\rho}(\mathbf{r})} \frac{\partial \tilde{\rho}(\mathbf{r})}{\partial D_{\mu\nu}^{\alpha}} d\mathbf{r} + \frac{\partial \left[E^{\alpha} - \tilde{E}^{\alpha}\right]}{\partial D_{\mu\nu}^{\alpha}}\right\} \frac{\delta D_{\mu\nu}^{\alpha}}{\delta \tilde{\phi}_i^*(\mathbf{r})} \\
&= f_i \left[-\frac{1}{2}\nabla_i^2 + \tilde{v}_{\mathrm{coul}}(\mathbf{r}) + \tilde{v}_{\mathrm{xc}}(\mathbf{r})\right] \tilde{\phi}_i(\mathbf{r}) \\
&\quad + \sum_{\alpha=1}^{N_{\mathrm{n}}} \sum_{\mu,\nu} \left\{\int \tilde{v}_{\mathrm{coul}}(\mathbf{r}) \frac{\partial \tilde{\rho}(\mathbf{r})}{\partial D_{\mu\nu}^{\alpha}} d\mathbf{r} + \frac{\partial \left[E^{\alpha} - \tilde{E}^{\alpha}\right]}{\partial D_{\mu\nu}^{\alpha}}\right\} f_i \tilde{p}_{\mu}^{\alpha}(\mathbf{r}) \langle \tilde{p}_{\nu}^{\alpha}|\tilde{\phi}_i\rangle \\
&= f_i \left[-\frac{1}{2}\nabla_i^2 + \tilde{v}_{\mathrm{coul}}(\mathbf{r}) + \tilde{v}_{\mathrm{xc}}(\mathbf{r})\right] \tilde{\phi}_i(\mathbf{r}) + \sum_{\alpha=1}^{N_{\mathrm{n}}} \sum_{\mu,\nu} f_i \tilde{p}_{\mu}^{\alpha}(\mathbf{r}) \Delta h_{\mu\nu}^{\alpha} \langle \tilde{p}_{\nu}^{\alpha}|\tilde{\phi}_i\rangle
\end{aligned}
$$

$$(5.91)$$

---

[1] See Appendix A of Ref. [3] for an overview of functional derivatives. In particular, the equations that are used here are Eqs. (A.24), (A.33), (A.34) of Ref. [3].

where, in going from line three to line four, we have further used that $\dfrac{\delta \tilde{\rho}(\mathbf{r}'')}{\delta \tilde{n}(\mathbf{r}')} =$

$\delta(\mathbf{r}'' - \mathbf{r}')$ and $\dfrac{\delta \tilde{n}(\mathbf{r}')}{\delta \tilde{\phi}_i^*(\mathbf{r})} = \delta(\mathbf{r}' - \mathbf{r})\tilde{\phi}_i(\mathbf{r})$. The Coulomb potential $\tilde{v}_{\mathrm{coul}}(\mathbf{r})$, and the xc

potential $\tilde{v}_{\mathrm{xc}}(\mathbf{r})$ are defined as:

$$\tilde{v}_{\mathrm{coul}}(\mathbf{r}) = \frac{\delta E_{\mathrm{coul}}\left[\tilde{\rho}\right]}{\delta \tilde{\rho}(\mathbf{r})} \tag{5.92}$$

$$\tilde{v}_{\mathrm{xc}}(\mathbf{r}) = \frac{\delta E_{\mathrm{xc}}\left[\tilde{n}\right]}{\delta \tilde{n}(\mathbf{r})} \tag{5.93}$$

and:

$$\Delta h_{\mu\nu}^{\alpha} = \int \tilde{v}_{\mathrm{coul}}(\mathbf{r}) \frac{\partial \tilde{\rho}(\mathbf{r})}{\partial D_{\mu\nu}^{\alpha}} d(\mathbf{r}) + \frac{\partial \left[E^{\alpha} - \tilde{E}^{\alpha}\right]}{\partial D_{\mu\nu}^{\alpha}} \tag{5.94}$$

By comparing Eq. (5.90) with the last line of Eq. (5.91), we obtain the explicit expression of the transformed KS Hamiltonian $\tilde{\mathbf{h}}_{\mathrm{KS}}$:

$$\tilde{\mathbf{h}}_{\mathrm{KS}} = -\frac{1}{2}\nabla_i^2 + \tilde{v}_{\mathrm{coul}}(\mathbf{r}) + \tilde{v}_{\mathrm{xc}}(\mathbf{r}) + \sum_{\alpha=1}^{N_{\mathrm{n}}} \sum_{\mu,\nu} |\tilde{p}_{\mu}^{\alpha}\rangle \Delta h_{\mu\nu}^{\alpha} \langle \tilde{p}_{\nu}^{\alpha}| \tag{5.95}$$

$\tilde{\mathbf{h}}_{\mathrm{KS}}$ is composed of three parts. The first part (first term on the right hand side of Eq. (5.95)) is a the kinetic energy operator. The second part ($\tilde{v}_{\mathrm{coul}}(\mathbf{r}) + \tilde{v}_{\mathrm{xc}}(\mathbf{r})$) represents an effective potential, and is a functional of only smooth pseudo densities. The third part (last term on the right hand side of Eq. (5.95)) is a correction term. As implied by Eq. (5.94), this correction is not just a constant potential, but adjusts together with the effective potential during the SCF steps.

5.4.5   GPAW: A Grid-Based Implementation of PAW

Inspired by already existing electronic structure codes based on pseudo-potentials, early implementations of the PAW method employed plane waves as basis in which to expand the pseudo orbitals [11, 12]. The GPAW program [23, 24] pursues a different strategy by representing orbitals, densities and potentials on real-space grids. The advantage of using real-space grids is twofold: first of all, systematic convergence of the accuracy of the representation is ensured by increasing the number of grid points per fixed volume (which is to say reducing the grid spacing); and, secondly, parallelization strategies based on efficient domain decomposition of the real-space grid, within the simulation box, can be adopted. Thanks to the latter, in particular, simulation times for large scale calculations can be significantly reduced when using parallel supercomputing systems [24, 26]. Moreover, GPAW takes advantage of the property of the pseudo orbitals and pseudo densities of being smooth everywhere in space, by representing them on relatively coarse grids. All atom-centered localized functions, such as the atomic partial waves, the projector functions and the core densities, are evaluated, instead, ahead of the actual calculations and stored in atomic setups. This allows to keep the memory requirements low and to boost even more the computational efficiency.

A representation of the electron density on a coarse real-space grid is particularly well-suited for multiscale embedding schemes that explicitly compute electrostatic interactions between the density of a solute and classical point change models representing the solvent. In Chap. 6, we will see how we have taken advantage of this computational expediency of GPAW to develop a QM/MM electrostatic embedding scheme [25] with only small added computational cost with respect to pure GPAW calculations of the isolated QM solute. Nonetheless, the multiscale strategy does not introduce approximations other than those already shared by standard implementations of QM/MM electrostatic embedding [25].

In GPAW all formulas to evaluate densities, potentials and energies are converted into discretized forms. For example, the Coulomb energy of Eq. (5.80) is computed as:

$$E_{\text{coul}}\left[\tilde{\rho}\right] = \frac{1}{2} V_g \sum_g \tilde{v}_{\text{coul}}(\mathbf{r}_g) \tilde{\rho}(\mathbf{r}_g) \tag{5.96}$$

where the summation is over points g of a uniform real-space grid, V_g is the volume per grid point, and the Coulomb potential $\tilde{v}_{\text{coul}}(\mathbf{r}_g)$ is obtained by solving a discretized version of the Poisson equation $\nabla^2 \tilde{v}_{\text{coul}}(\mathbf{r}) = -4\pi \tilde{\rho}(\mathbf{r})$. In essence, all integrals and derivatives are calculated using finite-difference methods. Iterative diagonalization schemes are, on the other hand, required to solve the generalized eigenvalue problem of Eq. (5.89).

In addition to the grid-based representation, linear combination of atomic orbitals (LCAO) basis sets are also available in GPAW [27] for representing the pseudo orbitals (when used, the densities and electrostatic interactions are still evaluated on the grid). The disadvantage of using LCAO basis sets is that converge of the accuracy, as in most LCAO-based electronic structure calculations [1, 24], cannot be reached as systematically as when using grid-based representations. However, there is an important advantage: the dimensionality of the problem when using LCAO basis is reduced with respect to the grid-based representation. This means that the memory requirements are even lower and that, for example, it is possible to solve the KS equations using direct diagonalization procedures. As a result, convergence of the SCF cycle is achieved more rapidly, and calculations with LCAO basis can be much faster than using finite-difference methods, for large systems. The accuracy of structural predictions of LCAO calculations in GPAW were found to be comparable to that of pure grid-based calculations [27]. As we will see later, SCF solution of the

KS equations represents the most computationally demanding part of direct QM/MM BOMD simulations. Since it is important to keep the computational cost at a minimum in order to collect statistical data sufficient to reach unequivocal conclusions about solution equilibrium and dynamical properties, all QM/MM BOMD simulations performed in the present work made use of GPAW with LCAO basis sets.

A last aspect of the program that is worth mentioning here, is that, in contrast to most electronic structure codes, which are based entirely on compiled languages like Fortran or C [26], GPAW adopts a Python/C combined approach [26]. The most computationally expensive operations, like matrix diagonalizations and operations on the grid, are carried out in C, but most of the program (about 85–90% of it [28])is actually written in Python.[2] This is done without significant speed loss, through extensive use of NumPy[3] for handling large arrays and communicating with C parts. Thanks to the high degree of modularity of object oriented programming in Python, it is relatively easy for users to add additional features in the code, as we have done for the ΔSCF implementation presented in Sect. 5.5.

5.5 Density Functional Theory for Excited States

The KS DFT formalism, as described in the previous sections, applies to electronic ground states. Generalization to the energetically lowest excited state of each symmetry (for symmetry we intend both the spatial symmetry, given by the irreducible representation of the point group, and the spin multiplicity of a state) is possible [29]. Strategies to solve the KS equations for single-determinant excited states variationally will be the topic of this section.

In the original KS scheme, only the N_e lowest energy orbitals that can be obtained from the eigenvalue problem of Eq. (5.27) are used to compute the (exact for a non-interacting system) expectation value of the kinetic energy operator and the electron density, according to Eqs. (5.20) and (5.21). Let us rewrite Eqs. (5.20) and (5.21) using an arbitrary number of spatial orbitals N_{orb}:

$$T_e^s[n] = \sum_{i=1}^{N_{orb}} f_i \langle \phi_i | -\frac{1}{2}\nabla_i^2 | \phi_i \rangle \tag{5.97}$$

$$n(\mathbf{r}) = \sum_{i=1}^{N_{orb}} f_i \mid \phi_i(\mathbf{r}) \mid^2 \tag{5.98}$$

---

[2]https://www.python.org/.

[3]http://www.numpy.org/.

In a ground-state calculation, one assigns the f_i according to the aufbau principle. Excited-state single determinants can be constructed by enforcing different occupations of the KS orbitals, through involvement of ground-state virtual (empty) orbitals. Note that the occupation numbers need not to be integers, in principle, but just those that guarantee the symmetry of the desired excited-state [29]. The KS equations are, then, solved variationally for the set of orbitals with constrained occupations. The procedure is known as ΔSCF [29, 30].

Originally, this term was used to indicate only computations of vertical transition energies by the difference between the energies of variationally optimized single-determinant excited and ground states, obtained at the same nuclear geometry. As the range of applications of the method has widened, the expression "ΔSCF" is now used to refer, more broadly, to any kind of DFT calculation that involves SCF convergence of a system using constraints on the occupation numbers. These include excited-state geometry optimizations [31], vibrational frequency calculations [32], PES scans [33–36], and BOMD simulations [37].

ΔSCF has been successfully employed to describe single-electron excitations, i.e. electronic excitations that, to a great extent, can be represented by the picture where one electron is promoted from an occupied to an empty orbital of the ground state, of a large variety of systems [38–44]. The performance of ΔSCF with respect to the prediction of vertical excitation energies was found to be comparable [39, 40] or, in some cases, even superior [38, 41] to that of time-dependent DFT (TDDFT), and the results are often in agreement with experiments and more advanced multireference wave function calculations [38, 42–44].

This success has prompted, in recent years, the development of practical solutions [34, 36, 45] to some of the deficiencies of ΔSCF, which limit its applicability in extended PES scans and BOMD simulations. Thus, for example, techniques like the maximum overlap method (MOM) [45], which avoid variational collapse to a lower state of the same symmetry during the SCF cycle, have been proposed, and find, nowadays, application in geometry optimizations [32] and BOMD simulations [37].

The increasing popularity of ΔSCF might seem surprising, given that the method lacks solid theoretical foundations, and, for this reason, its validity has been sometimes questioned [46]. Indeed, the Hohenberg-Kohn variational principle applies only to ground states, and there is no universal functional for excited states [46]. However, we must bear in mind that even for the ground state the variational principle is valid only when the *exact* functional is used [3]. Yet, ground-state DFT calculations employ, in practice, *approximate* functionals. Moreover, Van Voorhis et al. [40] have recently provided some theoretical justification to the use of ΔSCF, by showing that the method has a precise meaning within TDDFT with the adiabatic approximation.

ΔSCF is emerging as a cheap, yet accurate, alternative to TDDFT for structural predictions and BOMD simulations of the excited states of large systems, for which high-level multireference methods are not yet a viable choice. Preliminary investigations on small molecules [32, 36], organic dyes [34, 37] and even biological systems [33] are encouraging. In particular, ΔSCF was found to reproduce correct structures and PES topologies, even when the quality of excitation energies is inferior to that achieved by TDDFT [32] or higher-level methods [34]. Currently, there are no similar studies for transition metal complexes. One of the goals of the present work is to contribute to the understanding of the performance of ΔSCF by assessing the ability of the method to predict the structural dynamics of transition metal complexes.

Standard ΔSCF schemes based on promotion of a single electron from an occupied orbital $\phi_r(\mathbf{r})$ of the ground state to a virtual orbital $\phi_s(\mathbf{r})$, calculate the electron density of a system of N_e electrons as:

$$n(\mathbf{r}) = \sum_{i=1}^{N_e} (1 - \delta_{ri}) \, |\phi_i(\mathbf{r})|^2 + \sum_{j=N_e+1}^{N_{orb}} \delta_{sj} \, |\phi_j(\mathbf{r})|^2 \qquad (5.99)$$

where δ_{ri} and δ_{si} are delta functions. We will use the notation $|\Phi_r^s\rangle$ to indicate the excited state single determinant corresponding to the density given by Eq. (5.99).

5.5.1 Ziegler's Sum Method for Open-Shell Singlets

The lowest-lying singlet excited state of PtPOP is an open-shell singlet, since it possesses two unpaired electrons with opposite spin. There is an intrinsic limitation of the ΔSCF method, as illustrated until now, in treating open-shell systems as this one.

The single-determinant configurations that, intuitively, would seem to be the natural choice for describing an open-shell singlet with two unpaired electrons within ΔSCF, are represented schematically in Fig. 5.3. These single determinants, which we have indicated as $|\Phi_r^s\rangle$ and $|\Phi_{\bar{r}}^{\bar{s}}\rangle$, are not, however, pure singlet states. They have $M_S = 0$, but are not eigenfunctions of $\mathbf{S}^2$ [2, 4]. The expectation value of $\mathbf{S}^2$ with respect to either $|\Phi_r^s\rangle$ or $|\Phi_{\bar{r}}^{\bar{s}}\rangle$ is 1 [2], hence, the two wave functions can be considered as a mixture of singlet and triplet states. By taking appropriate linear combinations of $|\Phi_r^s\rangle$ and $|\Phi_{\bar{r}}^{\bar{s}}\rangle$ [2, 4], we can, however, construct a pure singlet:

$$|^1\Phi_r^s\rangle = \frac{1}{\sqrt{2}} \left(|\Phi_r^s\rangle + |\Phi_{\bar{r}}^{\bar{s}}\rangle \right) \qquad (5.100)$$

and a pure ($M_S = 0$) triplet states:

$$|^3\Phi_r^s\rangle = \frac{1}{\sqrt{2}} \left(|\Phi_r^s\rangle - |\Phi_{\bar{r}}^{\bar{s}}\rangle \right) \qquad (5.101)$$

Thus, we see that the open-shell singlet is more correctly described by the double determinant of Eq. (5.100). Obviously, ΔSCF cannot deal directly with $|^1\Phi_r^s\rangle$, because of its double-determinant character. We follow, instead, an indirect path, and combine Eqs. (5.100) and (5.101) to obtain an expression that relates $|\Phi_r^s\rangle$ to both $|^1\Phi_r^s\rangle$ and $|^3\Phi_r^s\rangle$:

$$|\Phi_r^s\rangle = \frac{1}{\sqrt{2}}\left(|^1\Phi_r^s\rangle + |^3\Phi_r^s\rangle\right) \tag{5.102}$$

By taking the expectation value of the electronic Hamiltonian with respect to $|\Phi_r^s\rangle$ as give by Eq. (5.102), and rearranging, we obtain:

$$\langle\Phi_r^s|\mathbf{H}_e|\Phi_r^s\rangle = \frac{1}{2}\left(\langle^1\Phi_r^s|\mathbf{H}_e|^1\Phi_r^s\rangle + \langle^3\Phi_r^s|\mathbf{H}_e|^3\Phi_r^s\rangle\right)$$
$$\Rightarrow \langle^1\Phi_r^s|\mathbf{H}_e|^1\Phi_r^s\rangle = 2\langle\Phi_r^s|\mathbf{H}_e|\Phi_r^s\rangle - \langle^3\Phi_r^s|\mathbf{H}_e|^3\Phi_r^s\rangle \tag{5.103}$$

where the singlet energy is now given in terms of the energies of the mixed spin state $|\Phi_r^s\rangle$ and the $M_S = 0$ triplet. This is the multiplet sum rule of Ziegler [30], which is usually written as:

$$E_S = 2E_M - E_T \tag{5.104}$$

The procedure consists in finding E_M and E_T from separate SCF optimization of single determinants. Since the $M_S = 0$ triplet is not a single determinant, in practice, E_T is obtained from the $M_S = 1$ triplet determinant $|^3\Phi_{\bar{r}}^s\rangle$ in an unrestricted calculation. In principle, the $M_S = 1$ and $M_S = 0$ triplets are degenerate. However, due to the approximate nature of the procedure, after the orbitals have relaxed in the SCF minimization, this is not strictly valid any more, and the use of $|^3\Phi_{\bar{r}}^s\rangle$ can be source of error in the determination of E_S.

Another inconvenience connected with the use of the sum rule, is that calculation of the gradients of the pure singlet state can be cumbersome, because it requires SCF convergence of two states, making geometry optimizations and BOMD simulations computationally expensive. Therefore, in the present work, we adopted a different strategy in the simulations of PtPOP in the S_1 state. Following Refs. [34, 39, 47], we computed the energy of the singlet open-shell from a single ΔSCF calculation of the restricted determinant corresponding to $|\Phi_r^s\rangle$ using the spin-unpolarized functional. Although spin-unpolarized ΔSCF calculations of open-shell singlets lack a formal theoretical foundation, their accuracy in estimating transition energies of transition metal complexes has turned out to be superior, in some cases, to the approach based on the sum rule [39]. This success was rationalized [39] on the basis of similarities between the spin-unpolarized ΔSCF density and an ensemble density [48].

5.5.2 Gaussian Smearing ΔSCF

Open-shell singlets are not the only systems whose multi-determinant character prevents application of the standard ΔSCF scheme exemplified by Eq. (5.99). Difficulties arise also when dealing with excitations that involve two or more degenerate orbitals.

As an example, let us consider, again, the CO molecule. Figure 5.4 shows the qualitative molecular orbital (MO) diagram of the ground state of CO, including the five highest occupied and three lowest unoccupied orbitals. The two lowest electronic excitations have $5\sigma \rightarrow 2\pi$ and $1\pi \rightarrow 2\pi$ character, respectively. Therefore, they both involve pairs of degenerate π orbitals. SCF convergence of a density obtained by changing the occupation number of only one of the two degenerate π orbitals by ± 1, according to Eq. (5.99), would be problematic, if not impossible at all. An *ad hoc* solution that is usually adopted in such cases, is to add (or remove) half electron to (from) both of the two degenerate π orbitals [31, 49].

However, this "trick" is not optimal for PES scans or BOMD simulations. In fact, the ordering of the orbital energies can change during the sampling, thus requiring a different occupation scheme for each nuclear configuration. In other words, what we need in order to perform PES calculations or BOMD simulations with ΔSCF without running into convergence problems, is a practical tool that allows to "dynamically" update the constraints on the occupation numbers.

Recently, Maurer et al. [34] have employed a modification of the standard ΔSCF constraints in ΔSCF PES calculations on azobenzene. The ordinary, discrete form of the ΔSCF constraints was replaced with Gaussian functions of the energies of the KS orbitals centered at the target orbitals ($\phi_r(\mathbf{r})$ and $\phi_s(\mathbf{r})$ in Eq. (5.99)). Such Gaussian smeared constraints affect all orbitals that lie close in energy to $\phi_r(\mathbf{r})$ and

Fig. 5.3 Schematic representation of two open-shell single determinants with mixed spin symmetry

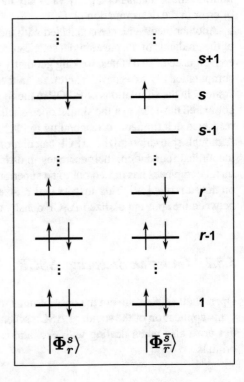

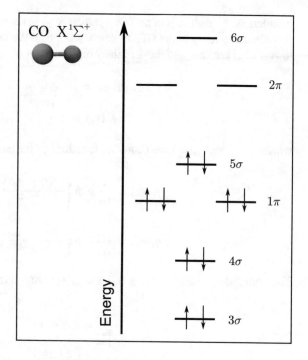

Fig. 5.4 Qualitative MO diagram of the CO molecule in the ground electronic state $(X^1\Sigma^+)$

$\phi_s(\mathbf{r})$, ensuring stable convergence of the density when state degeneracies are present, while avoiding smearing of the electrons for configurations in which the orbitals are well separated. The method has been demonstrated to be able to deliver, when applied to azobenzene using GGA functionals, PES topologies close to conical intersections (CIs) of quality comparable to those obtained using higher level Coupled Cluster Singles and Doubles calculations [34].

5.5.3 Implementing Gaussian Smearing ΔSCF in GPAW

The high density of states that characterizes transition metal complexes can be source of convergence issues in extensive ΔSCF excited-state QM/MM BOMD simulations. ΔSCF with Gaussian smeared constraints seems a promising strategy to ease the problem, due to its proven robustness and flexibility [34]. In order to investigate the possibility of using this tool in excited-state QM/MM BOMD simulations of systems like PtPOP, we have implemented it in a development branch of GPAW[4] [50].

As in Eq. (5.99), let r and s be indices for an occupied and a virtual orbitals of the ground state, respectively, and N_{orb} the total number of orbitals included in the

---

[4]The implementation is currently available within the following repository on Gitlab: https://gitlab.com/glevi/gpaw/tree/Dscf_gauss. The most relevant part of the code is included in Appendix A.

calculation. At each step of an SCF cycle, if N_{eval} is the number of valence electrons described explicitly in the GPAW calculation, then the occupation numbers of the i lowest N_{eval} orbitals, and those of the j orbitals from $N_{eval} + 1$ to N_{orb}, are calculated as:

$$f_i(\epsilon_i) = 1 - g_r(\epsilon_i) \tag{5.105}$$

$$f_j(\epsilon_j) = g_s(\epsilon_j) \tag{5.106}$$

where $g_r(\epsilon_i)$ and $g_s(\epsilon_j)$ are Gaussian functions of the energies of the KS orbitals:

$$g_r(\epsilon_i) = \frac{1}{N_r} \exp\left\{-\frac{(\epsilon_i - \epsilon_r)^2}{2\sigma^2}\right\} \tag{5.107}$$

$$g_s(\epsilon_j) = \frac{1}{N_s} \exp\left\{-\frac{(\epsilon_j - \epsilon_s)^2}{2\sigma^2}\right\} \tag{5.108}$$

The normalization factors for $g_r(\epsilon_i)$ and $g_s(\epsilon_j)$ are found by requiring that:

$$\sum_{i=1}^{N_{eval}} g_r(\epsilon_i) = 1 \tag{5.109}$$

$$\sum_{j=N_{eval}+1}^{N_{orb}} g_s(\epsilon_j) = 1 \tag{5.110}$$

such to satisfy a condition for conservation of the total number of electrons:

$$\sum_{i=1}^{N_{eval}}(1 - g_r(\epsilon_i)) + \sum_{j=N_{eval}+1}^{N_{orb}} g_s(\epsilon_j) = N_{eval} \tag{5.111}$$

The parameter σ controls the extent of the smearing, and can in principle be varied during the SCF cycle until satisfactory convergence is achieved.

Using the occupation numbers computed in this way, the modified form of the pseudo electron density (Eq. (5.60)) becomes:

$$\tilde{n}(\mathbf{r}) = \sum_{i=1}^{N_{eval}} f_i(\epsilon_i) \mid \tilde{\phi}_i(\mathbf{r}) \mid^2 + \sum_{j=N_{eval}+1}^{N_{orb}} f_j(\epsilon_j) \mid \tilde{\phi}_j(\mathbf{r}) \mid^2 + \tilde{n}_{core}(\mathbf{r}) \tag{5.112}$$

Also the elements $D_{\mu\nu}^{\alpha}$ of the atomic density matrix (Eq. (5.63)), and, therefore, all the atom-centered densities dependent on them, are changed by the constraints:

$$D_{\mu\nu}^{\alpha} = \sum_{i=1}^{N_{eval}} f_i(\epsilon_i)\langle\tilde{\phi}_i|\tilde{p}_{\mu}^{\alpha}\rangle\langle\tilde{p}_{\nu}^{\alpha}|\tilde{\phi}_i\rangle + \sum_{j=N_{eval}+1}^{N_{orb}} f_j(\epsilon_j)\langle\tilde{\phi}_j|\tilde{p}_{\mu}^{\alpha}\rangle\langle\tilde{p}_{\nu}^{\alpha}|\tilde{\phi}_j\rangle \quad (5.113)$$

Finally, the last expression that needs to be updated is that for the pseudo kinetic energy functional $T_e^s[\tilde{n}]$ (Eq. (5.66)):

$$T_e^s[\tilde{n}] = \sum_{i=1}^{N_{eval}} f_i(\epsilon_i)\langle\tilde{\phi}_i| -\frac{1}{2}\nabla_i^2|\tilde{\phi}_i\rangle + \sum_{j=N_{eval}+1}^{N_{orb}} f_j(\epsilon_j)\langle\tilde{\phi}_j| -\frac{1}{2}\nabla_j^2|\tilde{\phi}_j\rangle \quad (5.114)$$

All other expressions of the PAW formulation remain unaltered and the KS transformed equations can be solved, within GPAW, in the exact same way as illustrated in Sect. 5.4 for the ground state.

5.5.4 Testing the Implementation

We have tested our implementation of ΔSCF with Gaussian smeared constraints in GPAW with respect to the first two singlet and first two triplet excited states of the CO molecule. In what follows, we focus, in particular, on the performances with respect to structural predictions for the lowest-lying singlet states. The reason for this is that we aim at confidently applying the method in QM/MM BOMD simulations of the first singlet excited state of systems like PtPOP. Comparison of our calculations can be done with respect to two sets of results, reported in the literature, obtained with different implementations of ΔSCF [31, 36], as well as highly accurate experimental data [51–53].

The first implementation we compare to is the linear expansion ΔSCF (leΔSCF) method of Gavnholt et al. [36]. leΔSCF represents another variant of ordinary ΔSCF, in which electrons are added to (or removed from) linear combinations of KS orbitals. The method was already implemented in GPAW, and is tailored to study excitations of molecules adsorbed on metal surfaces. Handling degenerate π orbitals is not a problem within this approach, because the orbitals involved in the excitation can be taken as linear combinations of them [36], thus avoiding any convergence issue. We note, however, that leΔSCF is not suited for BOMD simulations, because it does not comply with the Hellman-Feynman theorem [36], the theorem that allows to compute analytical forces from the expectation value of the electronic Hamiltonian [54]. This is not a problem with our Gaussian smearing ΔSCF implementation, as we will see soon. Our second reference is a standard version of ΔSCF implemented in the DFT code CONQUEST [31]. In this case, fixed fractional occupation numbers were used for degenerate π orbitals when simulating the excited states of CO [31].

All calculations performed with our implementation of ΔSCF in GPAW employed an LDA xc functional. This choice is motivated by the fact that both reference calculations [31, 36] used this approximation. The width σ of the Gaussian functions (Eqs.

Table 5.1 Ground to excited state vertical excitation energies for the lowest singlet and triplet excited states of an isolated CO molecule computed at the equilibrium ground-state geometry using our GPAW implementation of ΔSCF with Gaussian smeared constraints. A comparison is made with calculated and experimental values retrieved from the literature. All values are in eV

State	Transition	Gaussian smearing ΔSCF GPAW		leΔSCF GPAW LDA[a] [36]	ΔSCF CONQUEST tzp/LDA [31]	Exp [51, 53]
		LDA[a]	tzp/LDA			
$A^1\Pi$	$5\sigma \rightarrow 2\pi$	7.82[b], 7.34[c]	7.71[b], 7.21[c]	7.84[b]	8.10[b]	8.51
$a^3\Pi$		6.09	5.93	6.09	5.26	6.32
$^1\Pi{-}^3\Pi$		1.73[b], 1.25[c]	1.78[b], 1.28[c]	1.75[b]	2.84[b]	2.19
$D^1\Delta$	$1\pi \rightarrow 2\pi$	10.75[b], 10.51[c]	10.65[b], 10.41[c]	10.82[b]	10.90[b]	10.23
$d^3\Delta$		9.66	9.54	9.72	9.11	9.36
$^1\Delta{-}^3\Delta$		1.09[b], 0.85[c]	1.11[b], 0.87[c]	1.10[b]	1.79[b]	0.87

[a] Grid-based representation of the orbitals
[b] Computed using Ziegler's sum rule [30]
[c] Obtained from spin-unpolarized calculations

(5.107) and (5.108)) controlling the extent of the smearing of the ΔSCF constraints was set to 0.01 eV. We have tested the implementation with both a pure grid-based representation of the orbitals, and using an LCAO tzp basis set [27]. The grid spacing was set to 0.18 Å, in any case. The grid-based calculations can be more closely compared to those performed using GPAW and the leΔSCF method, reported in Ref. [36], since the latter were also grid-based. The LCAO representation was tested because it can be used in QM/MM BOMD simulations of large systems in GPAW with considerable saving of computational cost, and is, therefore, the method of choice for such calculations.

Table 5.1 reports the vertical excitation energies for the two lowest singlet and triplet excited states of CO calculated at the ground-state optimized geometry by our GPAW implementation of Gaussian smearing ΔSCF, and the corresponding calculated and experimental reference values obtained from the literature [31, 36, 51, 53]. The singlet states ($A^1\Pi$ and $D^1\Delta$) are multi-determinant open-shell singlets (see paragraph Sect. 5.5.1). The calculations performed with leΔSCF in Ref. [36] and those realized with the program CONQUEST [31] used Ziegler's sum method to describe these states. For our tests, we report both the values obtained with the sum rule and those from a single calculation using the spin-unpolarized functional.

Overall, there is a satisfactory agreement between the transition energies computed with our implementation of Gaussian smearing ΔSCF and the values reported for the other two implementations of ΔSCF [31, 36]. Not surprisingly, the closest agreement is observed between the grid-based test calculations and the leΔSCF calculations in GPAW [36]. The LCAO tzp representation gives, in all cases, values that are only slightly smaller than those obtained with the pure grid technique. Use of the spin-unpolarized approximation for the singlets leads also to lower excitation energies as compared to the calculations that employed the sum method.

Table 5.2 Equilibrium bond lengths of CO in the ground state and lowest singlet and triplet excited states obtained with our implementation of Gaussian smearing ΔSCF in GPAW, and comparison with calculated and experimental values. All values are in Å

State	Transition	Gaussian smearing ΔSCF GPAW		ΔSCF CONQUEST tzp/LDA [31]	Exp [53]
		LDA[a]	tzp/LDA		
$X^1\Sigma^+$	Ground state	1.13	1.14	1.13	1.128
$A^1\Pi$	$5\sigma \rightarrow 2\pi$	1.21[b], 1.21[c]	1.23[b], 1.23[c]	1.22[b]	1.235
$a^3\Pi$		1.20	1.22	1.21	1.206
$D^1\Delta$	$1\pi \rightarrow 2\pi$	1.39[b], 1.36[c]	1.41[b], 1.39[c]	1.44[b]	1.399
$d^3\Delta$		1.36	1.39	1.38	1.370

[a] Grid-based representation of the orbitals
[b] Computed using Ziegler's sum rule [30]
[c] Obtained from spin-unpolarized calculations

The implementation of Gaussian smearing ΔSCF is intended to be used in extensive sampling of nuclear configurations in excited-state BOMD simulations. It is therefore important that the method is able to reproduce the shape of BO surfaces with sufficient accuracy over a wide range of configurations, independently of the absolute energy shift with respect to the ground state. Therefore, we have tested the performance of the implementation with respect to prediction of the PESs of CO in the lowest-lying excited states. A comparison can be made with experimental curves and equilibrium geometries, which are available [52, 53] from the analysis of highly resolved rovibrational spectra. As for reference calculations, we compare only to PESs computed using the CONQUEST implementation of ΔSCF [31], since PESs of CO obtained with leΔSCF in GPAW have not been reported.

Table 5.2 reports the equilibrium bond lengths of CO in the ground state and in the two lowest singlet and triplet excited states. The equilibrium geometries were obtained from geometry optimizations for all states except for the singlet excited states when described with Ziegler's sum rule; in these cases, the equilibrium bond lengths were extracted from the positions of the energy minima of the respective PESs, shown in Figs. 5.5(Right) and 5.6(Left).

Differences between the bond lengths optimized with our implementation of ΔSCF and the experimental values are all within 0.04 Å. In particular, for the $A^1\Pi$ excited states, despite differences between the computed excitation energies and experimental data as large as 1.3 eV (see Table 5.1), the equilibrium bond lengths deviate by less than 0.025 Å from experiments. More importantly, switching from the grid-based to the LCAO representation does not result in significant variations. This is in agreement with the finding that the LCAO description in GPAW tends to reproduce structural predictions of grid-based calculations very accurately, despite slightly larger errors, on average, for energies [27]. Analogously, the accuracy with respect to experiment does not seem to change substantially when using spin-unpolarized calculations for the open-shell singlets instead of the sum method.

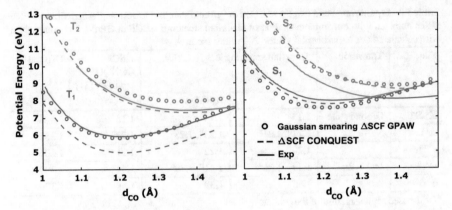

Fig. 5.5 Adiabatic PESs of CO in the lowest-lying excited states computed using our implementation of ΔSCF with Gaussian smeared constraints in GPAW, and comparison with curves obtained with another implementation of ΔSCF (digitalized from Ref. [31]) and determined from gas-phase rovibrational spectra (Ref. [52]). A grid-based representation was used for the orbitals. (Left) First two triplet states. (Right) First two singlet states calculated using Ziegler's sum rule

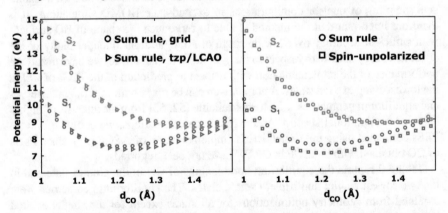

Fig. 5.6 Adiabatic PESs of CO in the two lowest-lying singlet excited states calculated using our GPAW implementation of ΔSCF with Gaussian smeared constraints. (Left) Comparison between the grid-based and the LCAO tzp calculations when using Ziegler's sum rule. (Right) Comparison between Ziegler's sum method and spin-unpolarized calculations when using a pure grid-based representation of the KS orbitals

Figures 5.5 and 5.6 show some of the adiabatic PESs of CO computed for the lowest-lying excited states using the Guassian smearing ΔSCF implementation in GPAW.

Comparison with the experimental curves [52, 53] and those computed by Terranova et al. [31] using ΔSCF in CONQUEST, confirms that all tested methods are able to reproduce the correct shapes of the PESs. In some cases, as for the T_1 state (see Fig. 5.5(Left)), the agreement with experiment of the GPAW ΔSCF calculations is improved with respect to ΔSCF in CONQUEST. Besides, for the singlet open-shells (Fig. 5.5(Right)) the calculations are able to reproduce the shapes of the experimental curves, but the position of the crossing between the S_1 and S_2 states is predicted at too large bond lengths. The origin of this discrepancy lies mainly in the error that affects the calculated energies for the $D^1\Delta$ (diabatic) state, which are too big compared to experiment. We should keep in mind, on the other hand, that the ΔSCF implementation is targeted, for the scopes of the present work, to applications on systems that do not exhibit strong deviation from the BO approximation. At this stage, an accurate prediction of conical intersections is beyond the ambitions of the method, especially for a diatomic molecule as CO, for which high-level multireference calculations are feasible. Figure 5.6(Left) shows that grid-based and LCAO calculations with a tzp basis set produce the same PESs for the singlet excited states, save for some small differences in the absolute positions of the minima. Finally, as seen from Fig. 5.6(Right), spin-unpolarized calculations are a valid alternative to Ziegler's sum method, as they virtually predict the same PESs (up to some constant shift).

Given the above results, we are confident that our implementation of ΔSCF with Gaussian smeared constraints in GPAW can be used for sufficiently reliable structural predictions in excited-state BOMD simulations. Moreover, the cost of QM/MM BOMD simulations of open-shell singlets can be kept to a minimum by using an LCAO representation of the orbitals, and the spin-unpolarized approximation, without losing accuracy.

Before concluding this section, we take a closer look at the ability of ΔSCF with Gaussian smeared constraints to deal with cases where state degeneracies would otherwise undermine stable converge of the SCF solutions.

We consider the first singlet excited states of CO. We attempted to compute the PESs in these states by replacing the smearing of the ΔSCF constraints with fixed, discrete constraints of the orbital occupation numbers. For the degenerate π orbitals the occupation numbers of the ground state were changed by ± 0.5. The calculations were performed in the spin-unpolarized approximation, and used a grid-based representation of the orbitals. Figure 5.7 shows the points on the PESs for which the SCF cycle could converge without problems. For S_2, almost all points could be converged. However, for bond lengths between ~ 1.4 and ~ 1.7 Å, the electronic density and orbitals of the S_1 (adiabatic) state could not be converged. Notably, convergence issues are experienced over a broad range of configurations, starting with points relatively far from the point where the two electronic states are expected to cross, around 1.6 Å.

Fig. 5.7 Points on the PESs
of an isolated CO molecule
in the first singlet excited
states obtained when using
ΔSCF in GPAW with
discrete constrains on the
orbital occupation numbers

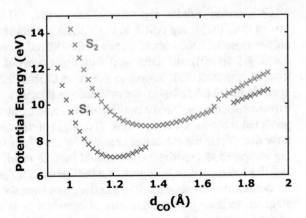

The example, although based on a simple diatomic system, is illustrative of the
challenges ΔSCF-QM/MM BOMD simulations in a lowest-lying singlet excited
state might face. Furthermore, one has to consider that the solvent can transiently
change the energy levels during the dynamics. As a result, state degeneracies could be
favoured even for configurations that would be energetically isolated in the gas-phase
system.

Smearing of the ΔSCF constraints is crucial in this regard. Figure 5.8 (Bottom)
shows that by using a Gaussian smearing with $\sigma = 0.01$ eV, as done in all calculations
presented before, it is possible to fully reconstruct the PES in the S_1 state, because
SCF convergence around the point of state crossing is no more a problem. The
top panels in Fig. 5.8 provide some insight into the issue and how it is overcome
by the method. Close to the minimum, the S_1 state has $5\sigma \rightarrow 2\pi$ character, the
5σ orbital has an occupation number of 1 and is relatively well separated from the
underlying, fully occupied 1π orbitals. For longer bond lengths, the energy difference
between the 5σ and 1π orbitals starts to decrease, until they are degenerate, around
1.6 Å. Gaussian smearing ensures stable convergence of the density at each point
by gradually changing the occupation numbers of the 5σ and 1π orbitals according
to their energy difference. At the point of crossing, the occupation numbers for all
three orbitals, the 5σ and two degenerate 1π orbitals, are basically the same. Note,
also, that due to the property of the Gaussian function of being peaked, it is not
until the energy difference between orbitals becomes smaller than ~0.04 eV, that the
smearing starts having an effect on the occupation numbers. At even longer bond
lengths, the energy of the 1π orbitals is higher than that of the 5σ orbital and the S_1
state has $1\pi \rightarrow 2\pi$ character.

Before, we have mentioned that ΔSCF implementations that involve linear combination of orbitals do not satisfy the Hellman-Feynman theorem [36], which is invoked
when computing analytical forces during geometry optimizations or BOMD simulations. We have tested whether Gaussian smearing of the orbital occupation num-

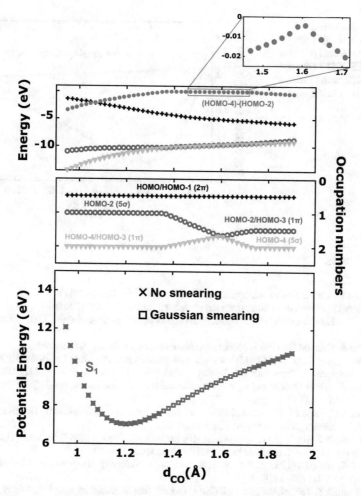

Fig. 5.8 (Bottom) Comparison between the points on the PES of CO in the S_1 excited state obtained using ordinary ΔSCF constrains (no smearing) and those calculated employing a Gaussian smearing of the constraints. (Middle) Occupation numbers of the five highest occupied KS orbitals along the CO bond length. (Top) Energies of the five highest occupied KS orbitals along the CO bond length

bers in our ΔSCF implementation can affect the quality of the analytical gradients. Figure 5.9 shows plots of the analytical gradients as computed for different points along the S_1 PES. Clearly, the gradients follow the slope of the S_1 curve, thus the smearing does not seem to be a limitation for analytical calculation of nuclear forces.

Fig. 5.9 Analytical
gradients computed by
GPAW at selected points on
the S_1 PES of CO obtained
using ΔSCF with Gaussian
smeared constraints

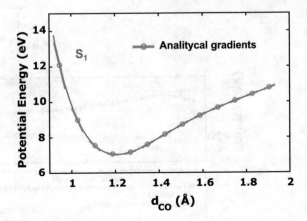

References

1. Jensen F Introduction to computational chemistry,3rd edn. Wiley (2017)
2. Cramer CJ Essentials of computational chemistry, 2nd edn. Wiley (2004)
3. Parr RG, Yang W (1989) Density-functional theory of atoms and molecules. Oxford University Press
4. Szabo A, Ostlund NS (1989) Modern quantum chemistry. Dover Publications
5. Hohenberg P, Kohn W (1964) Inhomogeneous electron gas. Phys Rev 136:B864–B871
6. Li L, Snyder JC, Pelaschier IM, Huang J, Niranjan U-N, Duncan P, Rupp M, Müller K-R, Burke K (2016) Understanding machine-learned density functionals. Int. J. Quantum Chem. 116(11):819–833
7. Snyder JC, Rupp M, Hansen K, Müller KR, Burke K (2012) Finding density functionals with machine learning. Phys Rev Lett 108:253002
8. Brockherde F, Vogt L, Li L, Tuckerman ME, Burke K, Müller KR (2017) Bypassing the Kohn-Sham equations with machine learning. Nat Commun 8(1):1–10
9. Kohn W, Sham LJ (1965) Self-consistent equations including exchange and correlation effects. Phys Rev 140:1133–1138
10. van Mourik T, Bühl M, Gaigeot M-P (2014) Density functional theory across chemistry, physics and biology. Philoso Trans R Soc Lond A Math Phys Eng Sci 372(2011):1–5
11. Blöchl PE, Först CJ, Schimpl J (2003) Projector augmented wave method: ab Initio molecular dynamics with full wave functions. Bull Mater Sci 26:33
12. Blöchl PE (1994) Projector augmented-wave method. Phys Rev B 50(24):17953
13. Perdew JP, Burke K, Ernzerhof M (1996) Generalized gradient approximation made simple. Phys Rev Lett 77:3865
14. Becke AD (1988) Density-functional exchange-energy approximation with correct asymptotic behavior. Phys Rev A 38:3098
15. Lee C, Yang W, Parr RG (1988) Development of the Colle-Salvetti correlation-energy formula into a functional of the electron density. Phys Rev B 37:785–789
16. Perdew JP (2013) Climbing the ladder of density functional approximations. MRS Bull 38(9):743–750
17. Vosko SH, Wilk L, Nusair M (1980) Accurate spin-dependent electron liquid correlation energies for local spin density calculations: a critical analysis. Can J Phys 58:1200–1211
18. Perdew JP, Wang Y (1992) Accurate and simple analytic representation of the electron-gas correlation energy. Phys Rev B 45:13244–13249
19. Becke AD (1986) Density functional calculations of molecular bond energies. J Chem Phys 84(8):4524–4529

20. Perdew JP, Chevary JA, Vosko SH, Jackson KA, Pederson MR, Singh DJ, Fiolhais C (1992) Atoms, molecules, solids, and surfaces: applications of the generalized gradient approximation for exchange and correlation. Phys Rev B 46:6671–6687

21. Becke AD (1993) Density-functional thermochemistry. III. The role of exact exchange. J Chem Phys 98:5648–5652

22. Stephens PJ, Devlin FJ, Chabalowski CF, Frisch MJ (1994) Ab-initio calculation of vibrational absorption and circular-dichroism spectra using density-functional force-fields. J Phys Chem 98:11623–11627

23. Mortensen JJ, Hansen LB, Jacobsen KW (2005) Real-space grid implementation of the projector augmented wave method. Phys Rev B 71:035109

24. Enkovaara J, Rostgaard C, Mortensen JJ, Chen J, Dulak M, Ferrighi L, Gavnholt J, Glinsvad C, Haikola V, Hansen HA, Kristoffersen HH, Kuisma M, Larsen AH, Lehtovaara L, Ljungberg M, Lopez-Acevedo O, Moses PG, Ojanen J, Olsen T, Petzold V, Romero NA, Stausholm-Møller J, Strange M, Tritsaris GA, Vanin M, Walter M, Hammer B, Häkkinen H, Madsen GKH, Nieminen RM, Nørskov JK, Puska M, Rantala TT, Schiøtz J, Thygesen KS, Jacobsen KW (2010) Electronic structure calculations with GPAW: a real-space implementation of the projector augmented-wave method. J Phys Condens Matter 22:253202

25. Dohn AO, Jónsson EÖ, Levi G, Mortensen JJ, Lopez-Acevedo O, Thygesen KS, Jacobsen KW, Ulstrup J, Henriksen NE, Møller KB, Jónsson H (2017) Grid-Based Projector Augmented Wave (GPAW) implementation of Quantum Mechanics/Molecular Mechanics (QM/MM) electrostatic embedding and application to a solvated diplatinum complex. J Chem Theory Comput 13(12):6010–6022

26. Enkovaara J, Romero NA, Shende S, Mortensen JJ (2011) GPAW—Massively parallel electronic structure calculations with Python-based software. Proc Comput Sci 4:17–25

27. Larsen AH, Vanin M, Mortensen JJ, Thygesen KS, Jacobsen KW (2009) Localized atomic basis set in the projector augmented wave method. Phys Rev B 80:195112

28. Larsen AH, Mortensen JJ, Blomqvist J, Castelli IE, Christensen R, Dułak M, Friis J, Groves MN, Hammer B, Hargus C, Hermes ED, Jennings PC, Jensen PB, Kermode J, Kitchin JR, Kolsbjerg EL, Kubal J, Kaasbjerg K, Lysgaard S, Maronsson JB, Maxson T, Olsen T, Pastewka L, Peterson A, Rostgaard C, Schiøtz J, Schütt O, Strange M, Thygesen KS, Vegge T, Vilhelmsen L, Walter M, Zeng Z, Jacobsen KW (2017) The atomic simulation environment–a python library for working with atoms. J Phys Condens Matter 29(27):273002

29. Gunnarsson O, Lundqvist BI (1976) Exchange and correlation in atoms, molecules, and solids by the spin-density-functional formalism. Phys Rev B 13(10):4274–4298

30. Ziegler T, Rauk A, Baerends EJ (1977) On the calculation of multiplet energies by the hartree-fock-slater method. Theor Chimica Acta 43(3):261–271

31. Terranova U, Bowler DR (2013) Δ self-consistent field method for natural anthocyanidin dyes. J Chem Theory Comput 9(7):3181–3188

32. Hanson-Heine MWD, George MW, Besley NA (2013) Calculating excited state properties using Kohn-Sham density functional theory. J Chem Phys 138(6):064101

33. Mendieta-Moreno J, Trabada DG, Mendieta J, Lewis JP, Gómez-Puertas P, Ortega J (2016) Quantum mechanics/molecular mechanics free energy maps and nonadiabatic simulations for a photochemical reaction in DNA: cyclobutane thymine dimer. J Phys Chem Lett 7(21):4391–4397

34. Maurer RJ, Reuter K (2011) Assessing computationally efficient isomerization dynamics: ΔSCF density-functional theory study of azobenzene molecular switching. J Chem Phys 135(22):1–25

35. Olsen T, Gavnholt J, Schiøtz J (2009) Hot-electron-mediated desorption rates calculated from excited-state potential energy surfaces. Phys Rev B 79(3):035403

36. Gavnholt J, Olsen T, Engelund M, Schiøtz J (2008) Delta self-consistent field as a method to obtain potential energy surfaces of excited molecules on surfaces. Phys Rev B 78:075441

37. Briggs EA, Besley NA, Robinson D (2013) QM/MM excited state molecular dynamics and fluorescence spectroscopy of BODIPY. J Phys Chem A 117(12):2644–2650

38. Briggs EA, Besley NA (2015) Density functional theory based analysis of photoinduced electron transfer in a Triazacryptand Based K+ sensor. J Phys Chem A 119:2902–2907

39. Himmetoglu B, Marchenko A, Dabo I, Cococcioni M (2012) Role of electronic localization in the phosphorescence of iridium sensitizing dyes. J Chem Phys 137(15):154309

40. Kowalczyk T, Yost SR, Van Voorhis T (2011) Assessment of the ΔSCF density functional theory approach for electronic excitations in organic dyes. J Chem Phys 134(5):054128

41. Robinson D, Besley NA (2010) Modelling the spectroscopy and dynamics of plastocyanin. Phys Chem Chem Phys 12(33):9667–9676

42. Fouqueau A, Mer S, Casida ME, Daku LML, Hauser A, Mineva T, Neese F (2004) Comparison of density functionals for energy and structural differences between the high- [5T2g:(t2g)4(eg)2] and low- [1A1g:(t2g)6(eg)2] spin states of the hexaquoferrous cation [Fe(H2O)6]2+. J Chem Phys 120(20):9473–9486

43. Johnson WTG, Sullivan MB, Cramer CJ (2001) Meta and para substitution effects on the electronic state energies and ring-expansion reactivities of phenylnitrenes. Int J Quantum Chem 85(4–5):492–508

44. Worthington SE, Cramer CJ (1997) Density functional calculations of the influence of substitution on singlet-triplet gaps in carbenes and vinylidenes. J Phys Organ Chem 10(10):755–767

45. Gilbert ATB, Besley NA, Gill PMW (2008) Self-consistent field calculations of excited states using the maximum overlap method (MOM). J Phys Chem A 112(50):13164–71

46. Gaudoin R, Burke K (2004) Lack of Hohenberg-Kohn theorem for excited states. Phys Rev Lett 93(17):1–4

47. Behler J, Reuter K, Scheffler M (2008) Nonadiabatic effects in the dissociation of oxygen molecules at the Al(111) surface. Phys Rev B Condens Matter Mater Phys 77(11):115421

48. Gross EKU, Oliveira LN, Kohn W (1988) Density-functional theory for ensembles of fractionally occupied states I. Basic formalism. Phys Rev A 37(8):2809–2820

49. Marshall D (2008) Computational studies of CO and CO+: density functional theory and time-dependent density functional theory. J Quant Spectrosc Radiat Transf 109(15):2546–2560

50. Levi G, Pápai M, Henriksen NE, Dohn AO, Møller KB (2018) Solution structure and ultrafast vibrational relaxation of the PtPOP complex revealed by ΔSCF-QM/MM direct dynamics simulations. J Phys Chem C 122:7100–7119

51. Nielsen ES, Jørgensen P, Oddershede J (1980) Transition moments and dynamic polarizabilities in a second order polarization propagator approach. J Chem Phys 73(12):6238

52. Tilford SG, Simmons JD (1972) Atlas of the observed absorption spectrum of carbon monoxide between 1060 and 1900 Å. J Phys Chem Ref Data 1(1):147–188

53. Huber KP, Herzberg G (1979) Molecular spectra and molecular structure, vol IV. Van Nostrand Reinhold, Constants of diatomic molecules

54. Feynman RP (1939) Forces in molecules. Phys Rev 56:340–343

Chapter 6
The Quantum Mechanics/Molecular Mechanics Method

Properly accounting for the influence of the solvent on the dynamics of transition metal complexes in BOMD simulations necessitates the use of atomistic models capable of describing explicit solute-solvent interactions. In fact, implicit solvation models, which represent the solvent as a continuum, are not able to describe, for example, specific transfer of excess vibrational energy from the solute to molecules of the solvent or solvent-induced vibrational dephasing. While modelling a system comprising a solute (in our case a transition metal complex) and an adequate number of solvent molecules with an electronic structure method like DFT can be impractical, one realizes that the solvent is amenable to less accurate, but computationally more expedient descriptions. As a matter of fact, processes like bond breaking/formation, or electronic excitations, which entail large electronic rearrangements, are usually confined within the solute (or within a solvation shell surrounding it, in the case, for example, of solute-solvent charge transfer reactions). Hybrid QM/MM methods divide the system of interest in a QM part, where the electronic structure is obtained at quantum mechanical level, and an MM part, where the level of treatment is based on molecular mechanics (MM), i.e. on classical potential functions. The idea is schematically illustrated in Fig. 6.1. Comprehensive reviews on development and application of QM/MM methodologies can be found, for example, in Refs. [2–4].

Different strategies exist for defining the boundary between the two regions [2, 3], whose level of complexity depends mainly on whether the QM/MM borders cut covalent bonds. Here, we will be concerned with nonadaptive QM/MM schemes, in which the partitioning in the two subsystems is kept fixed during a simulation, and the QM part includes the solute entirely.

Formally, the partition of the Hamiltonian and total energy of the full system, for an additive QM/MM scheme, can be expressed as:

$$\mathbf{H}_{TOT} = \mathbf{H}_{QM} + \mathbf{H}_{QM/MM} + \mathbf{H}_{MM} \tag{6.1}$$

$$E_{TOT} = E_{QM} + E_{QM/MM} + E_{MM} \tag{6.2}$$

Parts of this chapter have been reproduced with permission from Ref. [1]. Copyright 2018 American Chemical Society.

G. Levi, *Photoinduced Molecular Dynamics in Solution*, Springer Theses, https://doi.org/10.1007/978-3-030-28611-8_6

Fig. 6.1 Large systems that
need an explicit description,
like transition metal
complexes in solution, can be
simulated using a multiscale
QM/MM approach. The idea
behind it is to divide the
system into two parts (in our
case the solvated complex
and the solvent) based on the
different level of electronic
structure detail required by
each of them

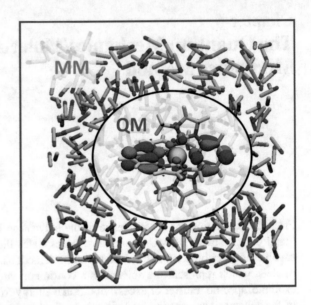

$\mathbf{H}_{QM}$ describes interactions between particles in the QM region, $\mathbf{H}_{MM}$ describes interactions between the classical MM particles, and $\mathbf{H}_{QM/MM}$ is a coupling Hamiltonian
accounting for interactions between QM and MM particles.

We already know $\mathbf{H}_{QM}$, because we have encountered it before in this thesis. In
general, $\mathbf{H}_{QM}$ has exactly the same form of the electronic Hamiltonian $\mathbf{H}_e$ of Eq.
(4.2).

Interactions between MM particles are represented with molecular mechanics
force fields, consisting in collections of classical pairwise additive potentials and
associated parameters. MM force fields usually describe a system of atoms with
point charges. There is not necessarily a one-to-one correspondence between MM
atoms and point charges, but charge sites can be displaced with respect to atomic
positions (in which case one can define "dummy" atoms carrying the charges) or
represent entire groups of atoms, as for example a methyl group. In addition there
can be Lennard-Jones (LJ) interaction sites, which account for dispersion and short-
range exchange repulsion (the van der Waals (vdW) interactions). For example, the
water TIP4P [5] model, which we employed in the QM/MM BOMD simulations of
PtPOP, consists of four interaction sites: two positive partial charges on the hydrogens
($q_H = 0.52$), one negative partial charge on a dummy atom M along the bisector of the
HOH angle ($q_M = -2q_H$), and a LJ site on the oxygen. Flexible MM models define
also an internal energy in terms of bonded potential functions. In this thesis, we will
be dealing only with force fields, like TIP4P, describing rigid solvent molecules.
For such force fields, the MM Hamiltonian (corresponding to the MM energy) of a
system of N_{MM} point charges and N_{LJ} LJ interaction sites is given by:

$$\mathbf{H}_{\text{MM}} = E_{\text{MM}} = \sum_{k=1}^{N_{\text{MM}}} \sum_{l>k}^{N_{\text{MM}}} \frac{q_k q_l}{|\mathbf{R}_k - \mathbf{R}_l|}$$

$$+ \sum_{\gamma=1}^{N_{\text{LJ}}} \sum_{\lambda>\gamma}^{N_{\text{LJ}}} 4\epsilon_{\gamma\lambda} \left[\left(\frac{\sigma_{\gamma\lambda}}{|\mathbf{R}_\gamma - \mathbf{R}_\lambda|} \right)^{12} - \left(\frac{\sigma_{\gamma\lambda}}{|\mathbf{R}_\gamma - \mathbf{R}_\lambda|} \right)^{6} \right] \quad (6.3)$$

where $\epsilon_{\gamma\lambda}$ and $\sigma_{\gamma\lambda}$ are the LJ parameters.

The major challenge connected with hybrid QM/MM methods is represented by the definition of the coupling Hamiltonian $\mathbf{H}_{\text{QM/MM}}$. Different levels of approximation can be adopted, ranging from mechanical embedding, in which neither of the two subsystems polarizes the other, to fully polarizable embedding. Without going into the details of each of them, in the following section, we will present only the so-called QM/MM electrostatic embedding scheme [6], in which only the QM subsystem is allowed to be polarized. The QM/MM electrostatic embedding scheme is implemented in GPAW [7], and has been used in the present work.

6.1 QM/MM Electrostatic Embedding

The electrostatic embedding QM/MM interaction Hamiltonian is defined as:

$$\mathbf{H}_{\text{QM/MM}}^{\text{el}} = -\sum_{k=1}^{N_{\text{MM}}} \sum_{i=1}^{N_e} \frac{q_k}{|\mathbf{r}_i - \mathbf{R}_k|} + \sum_{k=1}^{N_{\text{MM}}} \sum_{\alpha=1}^{N_n} \frac{q_k Z_\alpha}{|\mathbf{R}_\alpha - \mathbf{R}_k|} + \mathbf{H}_{\text{QM/MM}}^{\text{nb}}$$

$$= \mathbf{H}_{\text{QM/MM}}^{\text{coul}} + \mathbf{H}_{\text{QM/MM}}^{\text{nb}} \quad (6.4)$$

where the first and second terms account for the Coulomb interactions between electrons of the QM subsystem and MM charges, and between QM nuclei and MM charges, respectively. Other non-bonded (nb) terms include vdW interactions between QM and MM atoms, which are typically described with a LJ potential of the same form of the last term in Eq. (6.3):

$$\mathbf{H}_{\text{QM/MM}}^{\text{nb}} = E_{\text{QM/MM}}^{\text{nb}} = \sum_{\gamma=1}^{N_{\text{LJ}}} \sum_{\alpha=1}^{N_n} 4\epsilon_{\gamma\alpha} \left[\left(\frac{\sigma_{\gamma\alpha}}{|\mathbf{R}_\alpha - \mathbf{R}_\gamma|} \right)^{12} - \left(\frac{\sigma_{\gamma\alpha}}{|\mathbf{R}_\alpha - \mathbf{R}_\gamma|} \right)^{6} \right]$$

$$(6.5)$$

The Hamiltonian for the full system is:

$$\mathbf{H}_{\text{TOT}}^{\text{el}} = \mathbf{H}_{\text{QM}} + \mathbf{H}_{\text{QM/MM}}^{\text{el}} + \mathbf{H}_{\text{MM}} = \mathbf{H}_e + \mathbf{H}_{\text{QM/MM}}^{\text{el}} + \mathbf{H}_{\text{MM}} \quad (6.6)$$

where $\mathbf{H}_e$ is given by Eq. (4.2). The total energy of the *full* system can be obtained by solving the electronic Schrödinger equation (Eq. 4.3) for clamped QM nuclei and MM particles, in the same way as we would do to get the energy of an isolated QM system. Except that now the Hamiltonian is the Hamiltonian of the full system $\mathbf{H}_{TOT}^{el}$, and the wave function describing the QM subsystem has an additional parametric dependence on the positions of the MM particles. The first term on the right hand side of Eq. (6.4) that is included in $\mathbf{H}_{TOT}^{el}$ is entirely analogous to the $\mathbf{V}_{ne}$ term of the electronic Hamiltonian $\mathbf{H}_e$ (see Eq. 4.2), accounting for the Coulomb attraction between electrons and nuclei within the QM subsystem. That is to say, the external electrostatic potential of the MM charges acts on the electrons of the QM part in the same way as the "external" potential of the QM nuclei (Eq. 5.1) does. When solving the electronic Schrödinger equation with the Hamiltonian $\mathbf{H}_{TOT}^{el}$ defined in Eq. (6.6), using the variational principle, the wave function and the electron density of the QM subsystem will self-consistently relax with respect to the external potential of the MM charges. Thus, we see that, in the QM/MM electrostatic embedding scheme, the MM atoms are allowed to polarize the electron density of the QM part. All terms that do not have a dependence on the electronic coordinates of the QM subsystem, the last two terms on the right hand side of Eq. (6.4) and $\mathbf{H}_{MM}$, are constants for given QM and MM nuclear configurations, and, therefore, are similar to the nuclei-nuclei repulsion term $\mathbf{V}_{nn}$ in Eq. (4.2).

The expression for the total energy of the full system, Eq. (6.2), becomes:

$$E_{TOT} = E_{QM} + E_{QM/MM}^{el} + E_{MM} \tag{6.7}$$

where the total electrostatic embedding QM/MM interaction energy $E_{QM/MM}^{el}$ is given by the Coulomb interaction energy between MM and QM subsystems plus other QM/MM non-bonded (vdW) interactions:

$$E_{QM/MM}^{el} = E_{QM/MM}^{coul} + E_{QM/MM}^{nb} \tag{6.8}$$

The QM/MM electrostatic embedding formalism, as illustrated until now, is general, for any QM electronic structure method. Let us have a closer look at the particular case in which the QM subsystem is described using KS DFT.

In this case, one finds the KS orbitals of the Slater determinant that minimizes the energy of the full system by solving the KS equations (see Eq. 5.32) with a single particle KS Hamiltonian $\mathbf{h}_{KS}^{el}$, which includes the external electrostatic potential of the MM charges ($v_{ext}(\mathbf{r})$):

$$\mathbf{h}_{KS}^{el} = \mathbf{h}_{KS} + v_{ext}(\mathbf{r})$$
$$= -\frac{1}{2}\nabla_i^2 + v(\mathbf{r}) + v_H(\mathbf{r}) + v_{xc}(\mathbf{r}) + v_{ext}(\mathbf{r}) \tag{6.9}$$

$$v_{ext}(\mathbf{r}) = -\sum_{k=1}^{N_{MM}} \frac{q_k}{|\mathbf{r} - \mathbf{R}_k|} = \frac{\delta E_{QM/MM}^{coul}[n]}{\delta n(\mathbf{r})} \tag{6.10}$$

where the single-particle operators $v(\mathbf{r})$, $v_H(\mathbf{r})$, and $v_{xc}(\mathbf{r})$ have been defined in Eqs. (5.1), (5.30), and (5.31), respectively, and the functional $E_{QM/MM}^{coul}[n]$ for the electrostatic embedding QM/MM Coulomb interaction energy is:

$$E_{QM/MM}^{coul}[n] = -\sum_{k=1}^{N_{MM}} \int \frac{q_k n(\mathbf{r})}{|\mathbf{r} - \mathbf{R}_k|} d\mathbf{r} + \sum_{k=1}^{N_{MM}} \sum_{\alpha=1}^{N_n} \frac{q_k Z_\alpha}{|\mathbf{R}_\alpha - \mathbf{R}_k|} \quad (6.11)$$

Once the self-consistent minimization has produced converged orbitals and density, one can compute the total energy of the full system as:

$$\begin{aligned} E_{TOT} &= E_{KS} + E_{QM/MM}^{el} + E_{MM} \\ &= E_{KS} + E_{QM/MM}^{coul} + E_{QM/MM}^{nb} + E_{MM} \end{aligned} \quad (6.12)$$

where E_{KS} is the value of the energy functional defined in Eqs. (5.22) and (5.23), and $E_{QM/MM}^{nb}$ and E_{MM} are obtained from Eqs. (6.5) and (6.3), respectively.

6.1.1 QM/MM Electrostatic Embedding in GPAW

As specified in the introduction, the QM/MM BOMD simulations of PtPOP in water performed in the present work utilized the QM/MM electrostatic embedding method as implemented in the ASE [8, 9] and GPAW programs [7]. The implementation is the result of development work carried out in recent years with key contributions from the research group where this PhD project has been realized. The PhD student himself has been involved in the theoretical formulation of the method, and in the development of the routines to compute QM/MM LJ interactions. All details of the implementation have been presented in Ref. [7]. The code is available online within the official releases of ASE (https://gitlab.com/ase/ase) and GPAW (https://gitlab.com/gpaw/gpaw).

In this section, we provide the necessary information to understand how explicit QM/MM electrostatic interactions are computed using the PAW formulation of KS DFT. A broader overview of the QM/MM BOMD code will be given in the following section (Sect. 6.2).

As we have seen in Sect. 5.4 of Chap. 5, the PAW method works with pseudo orbitals $\tilde{\phi}_i(\mathbf{r})$ that replace the KS orbitals $\phi_i(\mathbf{r})$ in the SCF optimization of the electron density. This prevents straightforward addition of the external potential of the MM classical charges ($v_{ext}(\mathbf{r})$) to the PAW single-particle Hamiltonian $\tilde{\mathbf{h}}_{KS}$, defined in Eqs. (5.92)–(5.95). Instead, we need to evaluate a transformed operator $\tilde{v}_{ext}(\mathbf{r}) = \mathcal{T}^\dagger v_{ext}(\mathbf{r})\mathcal{T}$, where the PAW transformation operator $\mathcal{T}$ has been defined in Eq. (5.55). Analogously to the strategy employed in deriving the transformed KS Hamiltonian $\tilde{\mathbf{h}}_{KS}$ (see Eq. 5.90 in Sect. 5.4), we utilize the following relation:

$$\tilde{v}_{ext}(\mathbf{r})\tilde{\phi}_i(\mathbf{r}) = \frac{\delta E_{QM/MM}^{coul}[n]}{\delta \tilde{\phi}_i^*(\mathbf{r})} \tag{6.13}$$

We begin by writing the electrostatic embedding QM/MM Coulomb energy functional $E_{QM/MM}^{coul}[n]$ using the total charge density $\rho(\mathbf{r})$ defined in Eq. (5.71):

$$E_{QM/MM}^{coul}[n] = -\sum_{k=1}^{N_{MM}} \int \frac{q_k \rho(\mathbf{r})}{|\mathbf{r} - \mathbf{R}_k|} d\mathbf{r} \tag{6.14}$$

We can see that this energy functional is the same as in Eq. (6.11) by inserting the definition of $\rho(\mathbf{r})$ (contained in Eqs. 5.70 and 5.71) into Eq. (6.14):

$$E_{QM/MM}^{coul}[n] = -\sum_{k=1}^{N_{MM}} \int \frac{q_k \left(n(\mathbf{r}) - \sum_{\alpha=1}^{N_n} \delta(\mathbf{r} - \mathbf{R}) \mathcal{Z}_\alpha \right)}{|\mathbf{r} - \mathbf{R}_k|} d\mathbf{r}$$

$$= -\sum_{k=1}^{N_{MM}} \int \frac{q_k n(\mathbf{r})}{|\mathbf{r} - \mathbf{R}_k|} d\mathbf{r} + \sum_{k=1}^{N_{MM}} \sum_{\alpha=1}^{N_n} \frac{q_k \mathcal{Z}_\alpha}{|\mathbf{R}_\alpha - \mathbf{R}_k|} \tag{6.15}$$

Next, we rewrite Eq. (6.14) using the PAW definition of the electron density $n(\mathbf{r})$ given in Eq. (5.59):

$$E_{QM/MM}^{coul}[n] = -\sum_{k=1}^{N_{MM}} \int \frac{q_k \left(n(\mathbf{r}) - \sum_{\alpha=1}^{N_n} Z^\alpha(\mathbf{r}) \right)}{|\mathbf{r} - \mathbf{R}_k|} d\mathbf{r}$$

$$= -\sum_{k=1}^{N_{MM}} \int \frac{q_k \left[\tilde{n}(\mathbf{r}) + \sum_{\alpha=1}^{N_n} \tilde{Z}^\alpha(\mathbf{r}) + \sum_{\alpha=1}^{N_n} \left(n^\alpha(\mathbf{r}) + Z^\alpha(\mathbf{r}) - \tilde{n}^\alpha(\mathbf{r}) - \tilde{Z}^\alpha(\mathbf{r}) \right) \right]}{|\mathbf{r} - \mathbf{R}_k|} d\mathbf{r} \tag{6.16}$$

where we have further added and subtracted atom-centered compensation charges $\tilde{Z}^\alpha$. The compensation charges $\tilde{Z}^\alpha$ have been already introduced in Sect. 5.4 as functions that, by construction, make the electrostatic multipole moments of terms $n^\alpha(\mathbf{r}) + Z^\alpha(\mathbf{r})$ equal to those of terms $\tilde{n}^\alpha(\mathbf{r}) + \tilde{Z}^\alpha(\mathbf{r})$. The use of these functions leads to a simplification of the expression for the Coulomb interaction energy between electrons and nuclei of the QM subsystem, as explained in Sect. 5.4. Similarly, they also allow to achieve a simplification of the expression of the electrostatic embedding QM/MM Coulomb energy functional. In fact, by construction, terms $n^\alpha(\mathbf{r}) + Z^\alpha(\mathbf{r}) - \tilde{n}^\alpha(\mathbf{r}) - \tilde{Z}^\alpha(\mathbf{r})$ do not interact with the MM point charges q_k, and the electrostatic embedding QM/MM Coulomb interaction energy reduces to a functional of $\tilde{\rho}(\mathbf{r}) = \tilde{n}(\mathbf{r}) + \sum_{\alpha=1}^{N_n} \tilde{Z}^\alpha(\mathbf{r})$:

$$E_{\text{QM/MM}}^{\text{coul}}[n] = -\sum_{k=1}^{N_{\text{MM}}} \int \frac{q_k \left(\tilde{n}(\mathbf{r}) + \sum_{\alpha=1}^{N_{\text{n}}} \tilde{Z}^\alpha(\mathbf{r}) \right)}{|\mathbf{r} - \mathbf{R}_k|} d\mathbf{r}$$

$$= -\sum_{k=1}^{N_{\text{MM}}} \int \frac{q_k \tilde{\rho}(\mathbf{r})}{|\mathbf{r} - \mathbf{R}_k|} d\mathbf{r} = E_{\text{QM/MM}}^{\text{coul}}[\tilde{\rho}] \qquad (6.17)$$

Finally, with the definition of $E_{\text{QM/MM}}^{\text{coul}}[\tilde{\rho}]$ given in the second line of Eq. (6.17), the functional derivative $\frac{\delta E_{\text{QM/MM}}^{\text{coul}}[\tilde{\rho}]}{\delta \tilde{\phi}_i^*(\mathbf{r})}$ is evaluated using the same rules as employed to derive Eq. (5.91) in Sect. 5:

$$\frac{\delta E_{\text{QM/MM}}^{\text{coul}}[\tilde{\rho}]}{\delta \tilde{\phi}_i^*(\mathbf{r})} = \int \frac{\delta E_{\text{QM/MM}}^{\text{coul}}[\tilde{\rho}]}{\delta \tilde{n}(\mathbf{r}')} \frac{\delta \tilde{n}(\mathbf{r}')}{\delta \tilde{\phi}_i^*(\mathbf{r})} d\mathbf{r}' + \sum_{\alpha=1}^{N_{\text{n}}} \sum_{\mu,\nu} \frac{\partial E_{\text{QM/MM}}^{\text{coul}}[\tilde{\rho}]}{\partial D_{\mu\nu}^\alpha} \frac{\delta D_{\mu\nu}^\alpha}{\delta \tilde{\phi}_i^*(\mathbf{r})}$$

$$= \int \int \frac{\delta E_{\text{QM/MM}}^{\text{coul}}[\tilde{\rho}]}{\delta \tilde{\rho}(\mathbf{r}'')} \frac{\delta \tilde{\rho}(\mathbf{r}'')}{\delta \tilde{n}(\mathbf{r}')} \frac{\delta \tilde{n}(\mathbf{r}')}{\delta \tilde{\phi}_i^*(\mathbf{r})} d\mathbf{r}' d\mathbf{r}''$$

$$+ \sum_{\alpha=1}^{N_{\text{n}}} \sum_{\mu,\nu} \left[\int \frac{\delta E_{\text{QM/MM}}^{\text{coul}}[\tilde{\rho}]}{\delta \tilde{\rho}(\mathbf{r})} \frac{\partial \tilde{\rho}(\mathbf{r})}{\partial D_{\mu\nu}^\alpha} d\mathbf{r} \right] \frac{\delta D_{\mu\nu}^\alpha}{\delta \tilde{\phi}_i^*(\mathbf{r})}$$

$$= v_{\text{ext}}(\mathbf{r}) \tilde{\phi}_i(\mathbf{r}) + \sum_{\alpha=1}^{N_{\text{n}}} \sum_{\mu,\nu} \left[\int v_{\text{ext}}(\mathbf{r}) \frac{\partial \tilde{\rho}(\mathbf{r})}{\partial D_{\mu\nu}^\alpha} d\mathbf{r} \right] \tilde{p}_\mu^\alpha(\mathbf{r}) \langle \tilde{p}_\nu^\alpha | \tilde{\phi}_i \rangle$$

$$= v_{\text{ext}}(\mathbf{r}) \tilde{\phi}_i(\mathbf{r}) + \sum_{\alpha=1}^{N_{\text{n}}} \sum_{\mu,\nu} \tilde{p}_\mu^\alpha(\mathbf{r}) \Delta h_{\mu\nu}^{\alpha,\text{ext}} \langle \tilde{p}_\nu^\alpha | \tilde{\phi}_i \rangle \qquad (6.18)$$

where $\Delta h_{\mu\nu}^{\alpha,\text{ext}}$, appearing in the last line, is defined as:

$$\Delta h_{\mu\nu}^{\alpha,\text{ext}} = \int v_{\text{ext}}(\mathbf{r}) \frac{\partial \tilde{\rho}(\mathbf{r})}{\partial D_{\mu\nu}^\alpha} d(\mathbf{r}) \qquad (6.19)$$

By comparing the last line of Eq. (6.18) with Eq. (6.13) we derive the following expression for $\tilde{v}_{\text{ext}}(\mathbf{r})$:

$$\tilde{v}_{\text{ext}}(\mathbf{r}) = v_{\text{ext}}(\mathbf{r}) + \sum_{\alpha=1}^{N_{\text{n}}} \sum_{\mu,\nu} |\tilde{p}_\mu^\alpha\rangle \Delta h_{\mu\nu}^{\alpha,\text{ext}} \langle \tilde{p}_\nu^\alpha| \qquad (6.20)$$

Within the GPAW QM/MM electrostatic embedding scheme, the pseudo orbitals (and corresponding electron density) that minimize the total energy of the full system, are found by solving self-consistently the PAW transformed KS equations (see Eq. 5.89), with a single particle Hamiltonian consisting of the sum of the PAW Hamiltonian of Eq. (5.95) and the operator $\tilde{v}_{\text{ext}}(\mathbf{r})$:

$$\tilde{\mathbf{h}}_{KS}^{el} = \tilde{\mathbf{h}}_{KS} + \tilde{v}_{ext}(\mathbf{r}) \tag{6.21}$$

The total energy after convergence of the density can be computed from:

$$\begin{aligned} E_{TOT} &= E_{PAW} + E_{QM/MM}^{el} + E_{MM} \\ &= E_{PAW} + E_{QM/MM}^{coul} + E_{QM/MM}^{nb} + E_{MM} \end{aligned} \tag{6.22}$$

where E_{PAW} and $E_{QM/MM}^{coul}$ are obtained from Eqs. (5.83) and (6.17), respectively, $E_{QM/MM}^{nb}$ is computed from the LJ potential of Eq. (6.5), and E_{MM}, for a classical force field of rigid molecules, is given by Eq. (6.3).

We note that, in the derivation outlined above, we have not introduced approximations to the form of the QM/MM electrostatic embedding scheme. Besides, within GPAW, computation of QM/MM Coulomb interactions between the QM electronic density and the MM point charges (the most computationally demanding aspect of the QM/MM electrostatic embedding scheme) is straightforward and computationally efficient. Limiting the computational cost brought about by the calculation of the explicit QM/MM electrostatic interactions is achieved by exploiting the cost optimization tools inherent in GPAW [7, 10, 11]:

- Like all other potentials in GPAW, the external point charge MM potential is evaluated on domains of a real-space grid distributed among parallel processors (parallelization using domain decomposition).
- $\tilde{\rho}(\mathbf{r})$, which interacts with the MM charges in Eq. (6.17), is a smooth quantity and thus can be represented on relatively coarse grids.
- The size of the cell in which the QM subsystem is represented can be kept smaller than the cell of the full QM/MM system.

Lastly, a more technical consideration. The QM/MM electrostatic embedding scheme does not account for short-range exchange repulsion between electrons of the QM subsystem and MM atoms. This can cause an unphysical overpolarization of the QM electron density close to positive MM charges (the so-called charge spill-out effect [12]). To avoid this inconvenience, the implementation replaces, at distances below a certain cutoff radius, the basic form of the external potential (Eq. 6.10) with an analytical potential that goes smoothly towards a finite value for distances that tend to zero. The short-range potential has a 6th order polynomial form and matches the potential of Eq. (6.10) at the cutoff. More details about this aspect of the implementation can be found in Ref. [7].

6.2   Direct QM/MM Molecular Dynamics

In the previous chapters and sections, we have provided an overview of the strategies that, in the present work, have been adopted to find approximate solutions to the time-independent electronic Schrödinger equation for a system in either the ground or an excited state. In particular, we have focused on the GPAW DFT code, our ΔSCF implementation, and on how to define a hybrid solute-solvent system within it, for multiscale simulations. The last recipe we require to provide a complete view on the QM/MM BOMD strategy, as developed and employed in this project, is how to perform the classical propagation of such hybrid systems.

Newton's classical equations of motion have already been introduced in Sect. 4.2 of Chap. 4, in the context of fully ab initio BOMD simulations. Let us rewrite them here for a collection of atoms defining a QM/MM system:

$$\ddot{\mathbf{R}}_a - \frac{\mathbf{F}_a}{M_a} = 0 \tag{6.23}$$

where $\mathbf{R}_a$ and M_a are, respectively, the position vector and mass of particle a, which can be either a QM nucleus α or an MM atom k, and $\mathbf{F}_a$ is the force acting on it. In Eq. (6.23) we have used the notation $\dot{f} = \frac{\partial f}{\partial t}$ to indicate derivatives with respect to time.

The numerical integrator that is usually employed to solve Eq. (6.23) is the velocity Verlet algorithm [13], in which positions and velocities of the particles are propagated according to the following equations:

$$\mathbf{R}_a(t + \Delta t) = \mathbf{R}_a(t) + \dot{\mathbf{R}}_a(t)\Delta t + \frac{\mathbf{F}_a(t)}{2M_a}\Delta t^2 \tag{6.24}$$

$$\dot{\mathbf{R}}_a(t + \Delta t) = \dot{\mathbf{R}}_a(t) + \frac{\mathbf{F}_a(t) + \mathbf{F}_a(t + \Delta t)}{2M_a}\Delta t \tag{6.25}$$

where Δt is the classical time step. Generally, velocity Verlet is preferred over more elaborate numerical integration schemes, like the Runge–Kutta methods, because it provides good long-term stability of the total energy by ensuring time reversibility [14].

Equations (6.24) and (6.25) conserve the total energy of the system and, thus, generate a microcanonical (NVE) ensemble. When the interest is in equilibrium thermal properties or in the dynamics of a molecule in a heat bath, as is the case for the QM/MM BOMD simulations presented in this work, it is more desirable to perform the propagation in a canonical (NVT) ensemble. A commonly employed method in these cases is Langevin dynamics [14], in which friction terms γ_a and random forces $\mathbf{F}_a^{\text{rand}}$ are added to Newton's equations of motion:

$$\ddot{\mathbf{R}}_a = -\gamma_a \dot{\mathbf{R}}_a + \frac{\mathbf{F}_a}{M_a} + \frac{\mathbf{F}_a^{\text{rand}}}{M_a} \tag{6.26}$$

At time t in the propagation, the random force on particle a is connected to the target temperature T through:

$$\mathbf{F}_a^{\text{rand}}(t) = \sqrt{2 M_a k_b T \gamma_a} \eta_a(t) \tag{6.27}$$

where k_b is the Boltzmann constant, and η_a is a Gaussian random process. The advantage of using Langevin dynamics over less sophisticated thermostat methods based on rescaling of the velocities, like the Berendsen temperature-coupling scheme [15], is that the former generates a true NVT ensemble with the correct fluctuations of properties, while the latter produces only correct thermal averages but incorrect fluctuations [16]. Of course, when the focus is on the detailed microscopic dynamics of a solute, the Langevin (stochastic) thermostat has to be applied only to the solvent, which is done by setting to 0 the friction γ_a for the atoms of the solute, which will then be propagated by Newton's equations of motion.

Our QM/MM BOMD implementation uses the MD routines available in ASE [8] for numerical integration of the Langevin equations of motion. ASE implementation of Langevin dynamics is based on a generalization of the velocity Verlet algorithm [17] that can be used together with RATTLE distance constraints [18]. This type of constraints utilize the method of Lagrange multipliers, and are required for the MM subsystem, if the latter is described with a force field that does not allow for internal motion. Furthermore, they can also be applied to fix bond lengths and bond angles (indirectly, by constraining the distance between atoms that are both bonded to a third one) involving hydrogen atoms within the QM part. Applying constraints to the degrees of freedom of hydrogen atoms is done to increase the integration time step, in order to achieve longer simulation times at the same computational cost.

The intramolecular forces $\mathbf{F}_a$ needed for the classical propagation of the nuclei are computed from the nuclear gradients of the total energy of the system. In our GPAW implementation of QM/MM electrostatic embedding, the forces on the QM atoms α are obtained by differentiating the expression of the total energy given by Eq. (6.22) with respect to the nuclear positions $\mathbf{R}_\alpha$:

$$
\begin{aligned}
\mathbf{F}_\alpha &= -\frac{\partial E_{\text{PAW}}}{\partial \mathbf{R}_\alpha} - \frac{\partial E_{\text{QM/MM}}^{\text{coul}}}{\partial \mathbf{R}_\alpha} - \frac{\partial E_{\text{QM/MM}}^{\text{nb}}}{\partial \mathbf{R}_\alpha} \\
&= \mathbf{F}_{\alpha,\text{PAW}} + \mathbf{F}_{\alpha,\text{QM/MM}}^{\text{coul}} + \mathbf{F}_{\alpha,\text{QM/MM}}^{\text{nb}}
\end{aligned}
\tag{6.28}
$$

where the first two terms are computed as Hellmann-Feynman forces plus contributions from the response of the KS orbitals to nuclear displacements (see Ref. [19] for a description of how forces are calculated in the PAW method), whereas $\mathbf{F}_{\alpha,\text{QM/MM}}^{\text{nb}}$ is simply the derivative with respect to $\mathbf{R}_\alpha$ of the LJ potential of Eq. (6.5). For the MM atoms k, instead, we have:

$$
\begin{aligned}
\mathbf{F}_k &= -\frac{\partial E_{\text{QM/MM}}^{\text{coul}}}{\partial \mathbf{R}_k} - \frac{\partial E_{\text{QM/MM}}^{\text{nb}}}{\partial \mathbf{R}_k} - \frac{\partial E_{\text{MM}}}{\partial \mathbf{R}_k} \\
&= \mathbf{F}_{k,\text{QM/MM}}^{\text{coul}} + \mathbf{F}_{k,\text{QM/MM}}^{\text{nb}} + \mathbf{F}_{k,\text{MM}}
\end{aligned}
\tag{6.29}
$$

The first term in Eq. (6.29) is the most computationally expensive of all three, since it represents the force on MM atom k due to interaction with the electron density of the QM part. $\mathbf{F}^{coul}_{k,QM/MM}$ involves the following integral:

$$\mathbf{F}^{coul}_{k,QM/MM} = -\frac{\partial E^{coul}_{QM/MM}}{\partial \mathbf{R}_k}$$

$$= -\int \frac{q_k \tilde{\rho}(\mathbf{r})}{|\mathbf{r} - \mathbf{R}_k|^2} \frac{\mathbf{r} - \mathbf{R}_k}{|\mathbf{r} - \mathbf{R}_k|} d\mathbf{r} = -\int \frac{q_k \tilde{\rho}(\mathbf{r})}{|\mathbf{r} - \mathbf{R}_k|^3} (\mathbf{r} - \mathbf{R}_k) d\mathbf{r} \quad (6.30)$$

The other two terms are straightforward to compute, as they are obtained as the derivative with respect to $\mathbf{R}_k$ of LJ potentials and the Coulomb interaction energy between point charges (see Eqs. 6.3 and 6.5). Note that in case the interaction sites of the MM model do not coincide with the MM atoms, one first computes the forces on the interaction sites and then distributes them to the MM atoms according to the relative positions between interaction sites and MM atoms [20].

When simulating PtPOP in TIP4P water molecules, we have found it necessary to include counterions in the simulation box to avoid formation of vortices of water molecules around the complex, which we observed in the trajectories unwrapped from periodic boundary conditions (PBCs). The formation of vortices has been attributed to the large negative charge of the complex and could be removed with the addition of the counterions. In order to avoid any interference of the counterions on the dynamics of the solute, we have implemented an additional spherical harmonic potential that can be applied to the counterions to restrain them to parts of the simulation box far from the QM subsystem. The position restraint (pr) potential has the following form:

$$v_{pr}(\mathbf{R}_c) = \begin{cases} \frac{1}{2} k_{pr} \left(d'_c - d_{pr} \right)^2 & \text{if } d'_c \leq d_{pr} \\ 0 & \text{if } d'_c > d_{pr} \end{cases} \quad (6.31)$$

where $\mathbf{R}_c$ is the position vector of counterion c, $d'_c = |\mathbf{R}_c - \mathbf{R}^{CQM}|$ is the distance of the counterion from the center of the QM cell (CQM), and d_{pr} and k_{pr} are a cutoff radius and harmonic force constant, respectively. The forces on the counterions due to the harmonic restraint potential are given by:

$$\mathbf{F}_{c,pr} = \begin{cases} -k_{pr} \left(1 - \frac{d_{pr}}{d'_c} \right) (\mathbf{R}_c - \mathbf{R}^{CQM}) & \text{if } d'_c \leq d_{pr} \\ 0 & \text{if } d'_c > d_{pr} \end{cases} \quad (6.32)$$

Thus, the restrained counterions experience a harmonic force inside a sphere of radius d_{pr} and centered at $\mathbf{R}^{CQM}$ that drives them outside this region, where $\mathbf{F}_{c,pr} = 0$.

In all QM/MM BOMD simulations performed in this work, PBCs were treated by translating solvent molecules with respect to the center of the QM cell to conform the minimum image convention [13]. Furthermore, all electrostatic interactions within the QM/MM simulation box were computed in toto. We found that this did not significantly affect the computational cost for boxes of sizes as those used in the simulations. We note, on the other hand, that long-range cutoff schemes for the electrostatic interactions are available in the current implementation of the GPAW QM/MM code [7], following recent development work.

6.2.1 Overview of the QM/MM BOMD Code

Figure 6.2 shows the basic algorithm underlying direct QM/MM BOMD simulations in ASE and GPAW. Like GPAW, ASE is also written in Python. Object-oriented programming in Python offers a high degree of modularity and interfacing between different parts of the code. ASE takes care of creating and handling an atomistic object defining the QM/MM system of atoms. The different energy terms and the forces on the particles are obtained by calling, from within ASE, an interface calculator to GPAW and internal ASE calculators equipped with force fields for the MM part. Additional ASE modules perform the remaining tasks: applying PBCs, integrating the classical equations of motion, enforcing geometry constraints, outputting data. An overview of the ASE and GPAW modules that are involved in a QM/MM BOMD simulation is provided in Fig. 6.2.

In a nutshell, a QM/MM BOMD simulation in ASE and GPAW involves the following:

1. Set up the initial conditions for the dynamics. This consists in the initialization of an atomistic object containing positions $(\mathbf{R}_a(0))$ and velocities $(\dot{\mathbf{R}}_a(0))$ at time zero for all atoms of the QM/MM system. This step is done exclusively within ASE.

2. Compute atomic forces and total energy of the QM/MM system. The calculations are steered by a QM/MM interfacer (the qmmm.py module in ASE, see algorithm in Fig. 6.2), which also takes care of applying PBCs. The interfacer communicates with two calculators: (i) an MM force field calculator built in ASE, which computes MM total energy E_{MM} and forces due to interactions between MM point charges $(\mathbf{F}_{k,MM})$, and (ii) GPAW. In GPAW, the following takes place:

   - Set up the external potential of the MM point charges $v_{ext}(\mathbf{r})$.
   - Solve self-consistently the PAW transformed KS equations (Eq. 5.89) with single-particle KS Hamiltonian including $v_{ext}(\mathbf{r})$. In the case of excited-state simulations, this step involves the application of ΔSCF constraints on the orbital occupation numbers using the implementation of Gaussian smearing ΔSCF described in Sect. 5.5 of Chap. 5.
   - Compute the PAW total energy E_{PAW} (Eq. 5.83), the electrostatic embedding QM/MM Coulomb interaction energy $E_{QM/MM}^{coul}$ (Eq. 6.17), and the corresponding forces on the QM nuclei $(\mathbf{F}_{\alpha,PAW}, \mathbf{F}_{\alpha,QM/MM}^{coul})$.
   - Compute the forces on MM atoms due to Coulomb interactions between the MM point charges and the converged electronic density of the QM subsystem (Eq. 6.30).

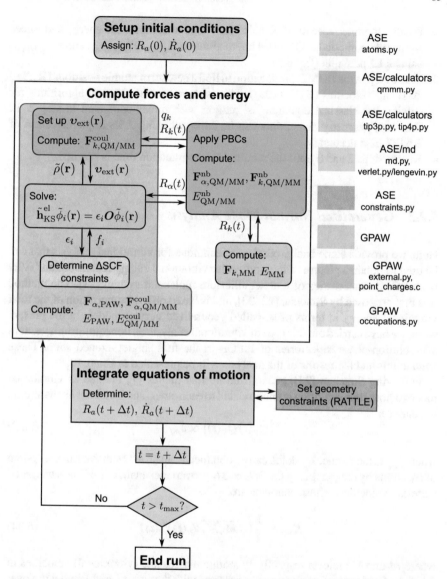

Fig. 6.2 Basic algorithm to perform on-the-fly QM/MM BOMD simulations in ASE & GPAW. The color code links ASE (https://gitlab.com/ase/ase) and GPAW (https://gitlab.com/gpaw/gpaw) modules to the specific tasks and operations they are called to fulfil during a simulation. The Gaussian smearing ΔSCF method has been implemented in the GPAW `occupations.py` module within the following development branch: https://gitlab.com/glevi/gpaw/tree/Dscf_gauss. Other modules that have been object of development work in the course of the Ph.D. project are the `constraints.py` and `qmmm.py` modules in ASE (available in the official release of the code)

Finally, the interfacer itself contains functions to compute the energy and forces from other non-bonded QM/MM interactions ($E_{QM/MM}^{nb}$, $\mathbf{F}_{\alpha,QM/MM}^{nb}$ and $\mathbf{F}_{k,QM/MM}^{nb}$) using a LJ potential (Eq. 6.5).

3. Take a step in the BOMD propagation to find a new set of atomic positions ($\mathbf{R}_a(t + \Delta t)$) and velocities ($\dot{\mathbf{R}}_a(t + \Delta t)$). This is done by ASE internal algorithms for solving the classical equations of motion with forces computed in step 2. In this step, geometry constraints can be enforced using ASE implementation of RATTLE, if necessary.
4. Repeat steps 2 and 3 until the required total simulation time is reached.

6.2.2   Generalized Normal Mode Analysis

Here, we provide some background on a technique for vibrational analysis that can be used to obtain a picture of intramolecular vibrational energy redistribution (IVR) from BOMD trajectories of a nonequilibrium molecular system [21]. The method was first proposed by Strachan [22, 23], and is based on a decomposition of the total vibrational energy in terms of so-called generalized normal modes. In the present work, we have carried out this type of vibrational analysis on nonequilibrium vacuum and solution-phase trajectories of PtPOP in the first singlet excited state. These simulations and the results of the analysis will be presented in Chap. 13.

Following Strachan [23], generalized normal modes Q_p defined as vibrational modes whose time evolution is uncorrelated to each other (and hence are not harmonic in general):

$$\langle \dot{Q}_p(t)\dot{Q}_q(t) \rangle \propto \delta_{pq} \tag{6.33}$$

where δ_{pq} is the Kronecker delta, can be obtained from an MD simulation of a system of N_n atoms by diagonalizing the $3N_n \times 3N_n$ covariance matrix $\mathbf{K}$ of mass weighted cartesian velocities, whose elements are:

$$K_{pq} = \frac{1}{2} \left\langle \sqrt{M_p M_q} V_p(t) V_q(t) \right\rangle \tag{6.34}$$

where M and V indicate respectively atomic masses and (vibrational) velocities in the body-fixed frame that translates and rotates with the system, and p and q run over the $3N_n$ cartesian components. The matrix $\mathbf{L}$ whose columns are the $3N_n$ normalized vibrational mode eigenvectors derived from diagonalization of $\mathbf{K}$ can be used to obtain a set of generalized normal mode velocities at each step of an MD trajectory by the following projection:

$$\dot{\mathbf{Q}}(t) = \mathbf{L}^T \mathbf{V}(t) \tag{6.35}$$

where $\dot{\mathbf{Q}}(t)$ and $\mathbf{V}(t)$ are $3N_n \times 1$ vectors of the instantaneous generalized normal mode and body-fixed-frame velocities, respectively, and $\mathbf{L}^T$ is the transpose of the matrix $\mathbf{L}$. The vibrational kinetic energy of the system can be decomposed into contributions from individual vibrational modes according to:

$$E_{\text{vib}}(t) = \frac{1}{2} \sum_{p=1}^{3N_n} \dot{Q}_p^2(t) = \sum_{p=1}^{3N_n} E_{\text{vib,p}}(t) \tag{6.36}$$

Thus, one can monitor the evolution of the portion of total vibrational kinetic energy shared by each generalized normal mode during a trajectory propagation, by projecting the body-fixed-frame velocities along the vibrational mode vectors through Eq. (6.35) and then computing the $E_{\text{vib,p}}(t)$ terms appearing in Eq. (6.36).

This procedure provides a means to draw a qualitative picture of intramolecular energy flow in a complex system and was recently successfully applied to analyse ab initio MD trajectories to investigate IVR processes in uracil [21]. Moreover, the generalized normal mode analysis briefly illustrated here was also used in another study, in conjunction with QM/MM simulations of a metal ion in water to decompose solute-solvent thermal fluctuations in terms of vibrational modes to support the analysis of X-ray absorption measurements [24].

We have implemented the method in a script using the Matlab programming language. The script is included in Appendix A.

References

1. Levi G, Pápai M, Henriksen NE, Dohn AO, Møller KB (2018) Solution structure and ultrafast vibrational relaxation of the PtPOP complex revealed by ΔSCF-QM/MM direct dynamics simulations. J Phys Chem C 122:7100–7119
2. Brunk E, Rothlisberger U (2015) Mixed quantum mechanical/molecular mechanical molecular dynamics simulations of biological systems in ground and electronically excited states. Chem Rev 115:6217–6263
3. Senn HM, Thiel W (2007) QM/MM methods for biological systems. Top Curr Chem 268:173
4. Warshel A, Levitt M (1976) Theoretical studies of enzymatic reactions: dielectric, electrostatic and steric stabilization of the carbonium ion in the reaction of lysozyme. J Mol Biol 103:227–249
5. Jorgensen WL (1981) Quantum and statistical mechanical studies of liquids. 10. transferable intermolecular potential functions for water, alcohols, and ethers. Application to liquid water. J Am Chem Soc 103:335–340
6. Field Martin J, Bash Paul A, Karplus Martin (1990) A combined quantum mechanical and molecular mechanical potential for molecular dynamics simulations. J Comput Chem 11(6):700–733
7. Dohn AO, Jónsson EÖ, Levi G, Mortensen JJ, Lopez-Acevedo O, Thygesen KS, Jacobsen KW, Ulstrup J, Henriksen NE, Møller KB, Jónsson H (2017) Grid-based projector augmented wave (GPAW) implementation of quantum mechanics/molecular mechanics (QM/MM) electrostatic embedding and application to a solvated diplatinum complex. J Chem Theory Comput 13(12):6010–6022

8. Larsen AH, Mortensen JJ, Blomqvist J, Castelli IE, Christensen R, Dułak M, Friis J, Groves MN, Hammer B, Hargus C, Hermes ED, Jennings PC, Jensen PB, Kermode J, Kitchin JR, Kolsbjerg EL, Kubal J, Kaasbjerg K, Lysgaard S, Maronsson JB, Maxson T, Olsen T, Pastewka L, Peterson A, Rostgaard C, Schiøtz J, Schütt O, Strange M, Thygesen KS, Vegge T, Vilhelmsen L, Walter M, Zeng Z, Jacobsen KW (2017) The atomic simulation environmenta python library for working with atoms. J Phys Condens Matter 29(27):273002
9. Bahn SR, Jacobsen KW (2002) An object-oriented scripting interface to a legacy electronic structure code. Comput Sci Eng 4:55
10. Enkovaara J, Romero NA, Shende S, Mortensen JJ (2011) GPAW-Massively parallel electronic structure calculations with Python-based software. Proc Comput Sci 4:17–25
11. Enkovaara J, Rostgaard C, Mortensen JJ, Chen J, Dulak M, Ferrighi L, Gavnholt J, Glinsvad C, Haikola V, Hansen HA, Kristoffersen HH, Kuisma M, Larsen AH, Lehtovaara L, Ljungberg M, Lopez-Acevedo O, Moses PG, Ojanen J, Olsen T, Petzold V, Romero NA, Stausholm-Møller J, Strange M, Tritsaris GA, Vanin M, Walter M, Hammer B, Häkkinen H, Madsen GKH, Nieminen RM, Nørskov JK, Puska M, Rantala TT, Schiøtz J, Thygesen KS, Jacobsen KW (2010) Electronic structure calculations with GPAW: a real-space implementation of the projector augmented-wave method. J Phys Condens Matter 22:253202
12. Laio A, VandeVondele J, Rothlisberger U (2002) A hamiltonian electrostatic coupling scheme for hybrid carparrinello molecular dynamics simulations. J Chem Phys 116:6941
13. Allen MP, Tildesley DJ (1989) Computer simulation of liquids. Oxford University Press
14. Jensen F (2017) Introduction to computational chemistry, 3rd edn. Wiley
15. Berendsen HJC, Postma JPM, van Gunsteren WF, DiNola A, Haak JR (1984) Molecular dynamics with coupling to an external bath. J Chem Phys 81(1984):3684–3690
16. Frenkel D, Smit B (2002) Understanding molecular simulation. Academic
17. Vanden-Eijnden E, Ciccotti G (2006) Second-order integrators for Langevin equations with holonomic constraints. Chem Phys Lett 429(1–3):310–316
18. Andersen HC (1983) Rattle: a "velocity" version of the shake algorithm for molecular dynamics calculations. J Comput Phys 52:24
19. Mortensen JJ, Hansen LB, Jacobsen KW (2005) Real-space grid implementation of the projector augmented wave method. Phys Rev B 71:035109
20. Feenstra KA, Hess B, Berendsen HJC (1999) Improving efficiency of large time-scale molecular dynamics simulations of hydrogen-rich systems. J Comput Chem 20(8):786–798
21. Carbonniere P, Pouchan C, Improta R (2015) Intramolecular vibrational redistribution in the non-radiative excited state decay of uracil in the gas phase: an ab initio molecular dynamics study. Phys Chem Chem Phys 17:11615–11626
22. Rega N (2006) Vibrational analysis beyond the harmonic regime from Ab-initio molecular dynamics. Theor Chem Acc 116:347–354
23. Strachan A (2004) Normal modes and frequencies from covariances in molecular dynamics or monte carlo simulations. J Chem Phys 120(1):1–4
24. Rega N, Brancato G, Petrone A, Caruso P, Barone V (2011) Vibrational analysis of x-ray absorption fine structure thermal factors by ab initio molecular dynamics: the Zn(II) ion in aqueous solution as a case study. J Chem Phys 134:074504

Part III
Time-Resolved Ultrafast X-Ray Scattering

Chapter 7
Observing Molecular Motion in Solution with X-Rays

Scattering of X-rays on a molecular system involves a change of the wave vector $|\mathbf{k}_0| = \dfrac{2\pi}{\lambda}$ of the incident photon with wavelength λ. This variation is described by the scattering vector $\mathbf{q}$:

$$\mathbf{q} = \mathbf{k}_0 - \mathbf{k}_s \tag{7.1}$$

where $\mathbf{k}_s$ is the wave vector of the scattered photon. In this thesis we will be concerned only with elastic scattering events, in which $|\mathbf{k}_s| = |\mathbf{k}_0|$, and:

$$q = |\mathbf{q}| = 2|\mathbf{k}_0| \sin\left(\frac{\theta}{2}\right) = \frac{4\pi}{\lambda} \sin\left(\frac{\theta}{2}\right) \tag{7.2}$$

where θ is the angle between the wave vectors $\mathbf{k}_s$ and $\mathbf{k}_0$. This fundamental process of light-matter interaction can be exploited for structural determinations at the atomic scale resolution. X-ray scattering techniques to obtain the structure of molecules in crystal or solution phase have undergone tremendous improvements over the last decades [1].

One of the greatest achievements reached in the X-ray field in recent years is the development of time-resolved X-ray scattering techniques to probe atomic motion as it occurs in real time [2–5]. The aim of the present chapter is to give a general overview of pump-probe X-ray scattering experiments as performed by the group of our experimental collaborators to investigate photoinduced dynamical processes in complex molecular systems in solution. Subsequently, in Chap. 8, we will see how these experiments are typically analysed in the group of our experimental collaborators, and how BOMD data can be used to simulate the scattering signal, offering assistance to the interpretation of the experimental outcomes.

© Springer Nature Switzerland AG 2019
G. Levi, *Photoinduced Molecular Dynamics in Solution*, Springer Theses,
https://doi.org/10.1007/978-3-030-28611-8_7

7.1   Pump-Probe Experiments at XFELs

Time-resolved structural determinations with X-rays are based on the pump-probe operative principle: an ultrashort optical pump laser initiates a reaction, which is probed, after a time delay t_p, by an ultrashort X-ray pulse focused on the sample. By collecting and putting in a sequence scattering images acquired at different pump-probe time delays, it is possible to produce a "molecular movie" of the dynamics consisting of a series of "snapshots" of atomic motion, if the setup permits to achieve an adequate time resolution. The pump-probe experiments that we perform use a UV-vis ultrashort laser that triggers nuclear dynamics by electronically exciting solute molecules in dilute solution. The use of a UV-vis pump pulse implies that the processes that can be studied with this technique are photochemically activated reactions. However, one should not think that this confines the investigation to excited-state PESs. As we will see in the next section, a careful choice of pump-pulse parameters can enable direct probing of ground-state PESs as well.

In a typical optical pump/X-ray probe experiment in solution, the beams of the two lasers are focused on a liquid jet with the sample. The liquid jet is produced through a nozzle and allows continuous replenishment of the sample [6]. The X-ray signal is collected, at a given pump-probe time delay, on a 2D detector positioned on a plane perpendicular to the direction of propagation of the X-ray beam. A schematic diagram of the setup of a standard time-resolved X-ray scattering experiment in solution is shown in Fig. 7.1.

Molecular reactions involving bond breaking/formation unfold on femtosecond time scales. Therefore the pump-probe apparatus that we require to investigate such reactions should ensure femtosecond time resolution. In ordinary X-ray diffraction measurements on crystals, the long-range order that characterizes the sample allows constructing interference of the diffracted waves for special $\mathbf{q}$ directions. As a

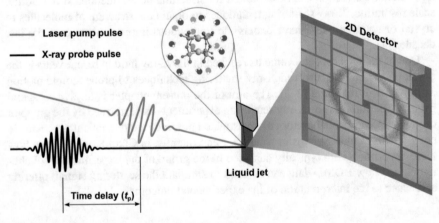

Fig. 7.1 Schematical illustration of an optical pump/X-ray probe setup for time-resolved X-ray diffuse scattering (XDS) experiments

consequence, the diffraction pattern features sharp peaks, the so-called Bragg peaks, at specific locations. Compared to ordered materials, disordered molecules in solution give rise to diffuse and weak diffraction patterns covering a vast portion of $\mathbf{q}$ space (as the illustration of a 2D scattering pattern of a solution in Fig. 7.1 suggests). This is due to the broad range of orientations and nuclear configurations that exist in solution. Commonly, one refers to scattering from disordered materials as "X-ray diffuse scattering" (XDS).

The time resolution of a pump-probe XDS experiment can be limited by either one of these two factors: (i) the shortest time in which the X-ray beam is able to deliver enough photons for a detectable signal to be measured, and (ii) the jitter between pump and probe pulses. The only coherent X-ray sources existing in the world that are able to produce hard X-ray pulses as short as a few fs, with sufficient brilliance, are X-ray free-electron lasers (XFELs) [2, 7–9]. The world's first XFEL, the Linac Coherent Light Source (LCLS)[1] of Stanford, became operational in 2009 [9]. The PtPOP time-resolved XDS measurements that will be presented in the next section have been performed at LCLS. The second oldest XFEL is SACLA[2], in Japan. Three other facilities started operation in 2017: the European XFEL[3] in Hamburg, the Pohang Accelerator Laboratory[4] in South Korea, and the SwissFEL at the Paul Scherrer Institute in Switzerland.[5] These large scale facilities provide flashes of X-rays with a duration of tens of fs [2, 7, 10], and can achieve fluxes of around 10^{12} photons per pulse [2, 7, 10]. Moreover, they are equipped with timing-tools to control with femtosecond resolution the jitter between the laser pump and the X-ray probe [2, 7, 11].

7.1.1 The Difference Scattering Signal

In order to enhance photoinduced structural changes, one usually records difference scattering signals during a time-resolved XDS experiment, by taking, at each time delay, the difference between an image with the laser pump pulse turned on ($S_{on}(\mathbf{q}, t_p)$) and one with the laser turned off ($S_{off}(\mathbf{q}, t_p)$):

$$\Delta S(\mathbf{q}, t_p) = S_{on}(\mathbf{q}, t_p) - S_{off}(\mathbf{q}, t_p) \tag{7.3}$$

This procedure removes all contributions to the scattering that are unaltered by photoexcitation, which include inelastic scattering [12] or the scattering from that large portion of the sample that is not perturbed by the laser.

---

[1]https://lcls.slac.stanford.edu/.

[2]http://xfel.riken.jp/eng/.

[3]https://www.xfel.eu/.

[4]http://pal.postech.ac.kr/paleng/.

[5]https://www.psi.ch/swissfel/.

The signal when the laser pump pulse is off is the scattering from a (stationary) thermal equilibrium ensemble of ground-state molecules ($S_{GS}^{eq}(\mathbf{q})$):

$$S_{off}(\mathbf{q}, t_p) = S_{GS}^{eq}(\mathbf{q}) \tag{7.4}$$

After laser excitation, the population of molecules in solution will be distributed over the electronic ground state and an excited state (assuming excitation occurs to a single electronic state). Hence, the scattering after the laser pump pulse has been on is made up of contributions from the non-stationary excited-state population ($S_{ES}(\mathbf{q}, t_p)$) and from the ensemble of molecules remaining in the ground state:

$$S_{on}(\mathbf{q}, t_p) = \alpha S_{ES}(\mathbf{q}, t_p) + S_{GS}^{eq}(\mathbf{q}) - \alpha S_{GS}^{h}(\mathbf{q}, t_p) \tag{7.5}$$

where α is the fraction of excited-state molecules, and $S_{GS}^{h}(\mathbf{q}, t_p)$ is the signal arising from the difference between the nuclear distribution of the equilibrium ground-state ensemble and that of the non-stationary ensemble of molecules remaining in the ground state after excitation. $S_{GS}^{h}(\mathbf{q}, t_p)$ is the signature of the hole left in the ground-state distribution by the pulse. α is assumed to be constant in time. We will only consider scattering signals recorded at times considerably shorter than the lifetime of the excited state, for which the assumption of a constant α is valid. By combining Eqs. (7.3), (7.4) and (7.5), we see that the difference scattering signal can be expressed as:

$$\Delta S(\mathbf{q}, t_p) = \alpha \left[S_{ES}(\mathbf{q}, t_p) - S_{GS}^{h}(\mathbf{q}, t_p) \right] \tag{7.6}$$

7.1.2 Anisotropy in Time-Resolved X-Ray Scattering Signals

Another important aspect of the scattering signal has to be mentioned. When using a linearly polarized laser pump pulse, as usually is the case in standard pump-probe setups, among the molecules that are oriented randomly in solution, the laser will preferentially excite those with the transition dipole moment parallel to the laser polarization axis. Therefore, for times shorter than the rotational correlation time of the photoexcited molecule in solution, the X-ray scattering signal will appear anisotropic. With the femtosecond time resolution available at XFELs, these time scales have become accessible, since rotational dephasing of aligned molecules in solution happens, usually, for times longer than 50 ps for medium-sized solutes [13].

In general, the theoretical interpretation of anisotropic scattering patterns of poly-atomic molecules is complicated [14]. However, the treatment considerably simpli-fies in the case of symmetric top molecules, like PtPOP, with the transition dipole moment parallel to the unique axis of symmetry, as it has been shown by Baskin and Zewail for ultrafast electron diffraction [15, 16], and later by U. Lorenz et al. in our group for X-ray scattering signals [14]. We further restrict out attention to one-photon absorption processes. If all the symmetry restrictions listed before are met, one-photon absorption of a linearly polarized pulse by a thermal molecular ensemble results in the formation of an excited-state population with a cosine-squared distribution of the dipole moment with respect to the laser polarization axis, and of a hole, representing depletion of the equilibrium ground-state ensemble, having the same rotational symmetry as the excited state [14, 15, 17]. In this case, the difference scattering signal can be decomposed into two contributions [14, 18]:

$$\Delta S(q, \theta_q, t_p) = \Delta S_0(q, t_p) + P_2(\cos \theta_q) \Delta S_2(q, t_p) \qquad (7.7)$$

where P_2 is the second-order Legendre polynomial ($P_2(x) = (3x^2 - 1)/2$) and θ_q is the angle between the laser polarization axis and the scattering vector $\mathbf{q}$, which can be inferred from the geometry of the experiment [14, 15]. Extracting $\Delta S_0(q, t_p)$ and $\Delta S_2(q, t_p)$ from the total difference scattering signal is a straightforward linear fitting procedure at each fixed value of q [14, 18]. The two terms can then be analysed separately. $\Delta S_0(q, t_p)$ is an isotropic term, in that it has the same form of the X-ray scattering signal one would obtain from an isotropic ensemble [14, 18], and can be analysed using the same tools that are used to interpret scattering patterns of isotropic samples. $\Delta S_2(q, t_p)$ is called anisotropic scattering term. It contains coefficients related to the rotational distribution of the photoexcited ensemble, and further encodes information on the orientation of atomic distances with respect to the transition dipole moment [14, 18].

Techniques to process anisotropic scattering signals according to the separation in the two contributions $\Delta S_0(q, t_p)$ and $\Delta S_2(q, t_p)$ have been refined and are becoming routine in the group of our experimental collaborators [18, 19]. Recently, we have carried out one of the first quantitative analysis of the anisotropic contribution $\Delta S_2(q, t_p)$ in the scattering signal of a photoexcited complex in solution [18]. The investigated complex was PtPOP, and the analysis allowed to obtain the value of the Pt–Pt distance in solution, which is the key structural parameter of the molecule. Part of the present Ph.D. project has been spent in assisting this analysis. Methods for simulating X-ray scattering signals to analyse the experimental data will be briefly discussed in Chap. 8. In the next section we will present the time-resolved XDS experiment on PtPOP that we have performed at the LCLS XFEL of Stanford.

7.2 Measuring PtPOP in Water

We have measured the time-dependent X-ray scattering signal upon photoexcitation by an ultrashort optical pulse of a dilute 80 mM aqueous solution of PtPOP with femtosecond time resolution. The measurements were performed at the X-ray Pump Probe (XPP) experimental station [8] at the LCLS XFEL facility of Stanford. In the experiment, a linearly polarized pump pulse with duration of ∼50 fs (full width at half maximum (FWHM) of the spectral intensity profile) and wavelength of ∼395 nm was used. Figure 7.2 shows a Gaussian fit to the spectral intensity profile of the pump pulse, together with the experimental absorption spectrum of PtPOP in water [20]. In the range of wavelengths shown in the figure, excitation occurs from the ground

Fig. 7.2 Gaussian fit (purple line) to the spectral intensity profile of the ultrashort pump pulse used in the time-resolved XDS experiment on PtPOP in water [18, 20], together with the $S_0 \to S_1$ band (black line) of the absorption spectrum of the molecule measured in water [20]

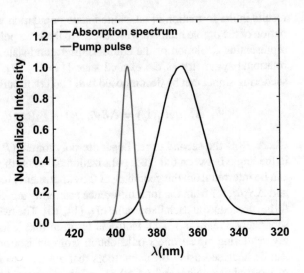

state (S_0) to the lowest-lying singlet excited state (S_1) of the complex. Figure 7.2 highlights that the pulse covered a range of excitation energies at the far red side of the 370 nm maximum of the $S_0 \to S_1$ absorption band.

The raw data collected at the detector were processed by T. B. van Driel to apply the corrections described in Ref. [21]. Afterwards, 2D difference scattering images were obtained by taking the difference between the corrected images acquired with the laser on and the corrected images acquired with the laser off. Finally, 1D isotropic and anisotropic difference scattering curves were obtained, at each pump-probe time delay, according to Eq. (7.7), following the procedure outlined in Ref. [18].

Figure 7.3 shows the isotropic and anisotropic contributions to the difference scattering signal, $\Delta S_0(q, t_p)$ and $\Delta S_2(q, t_p)$, as a function of the time delay t_p and the magnitude of the scattering vector $\mathbf{q}$. Both signals exhibit a pronounced beating pattern that lasts for at least ~3.5 ps. The two data sets were analysed by singular-value decomposition (SVD), and the Fourier transforms (FTs) of the first right-singular vectors of $\Delta S_0(q, t_p)$ and $\Delta S_2(q, t_p)$ are plotted in the insets of Fig. 7.3. The FTs reveal that the oscillations in the scattering signals have a period of ~285 fs. This value is very close to the period assigned to Pt–Pt stretching vibrations in the ground state from Raman spectroscopic (~283 fs) [22] and transient absorption (~281 fs) [23] measurements in aqueous solution. Besides, there is no clear peak in the FTs around 224 fs, which is the value of the Pt–Pt vibrational period characteristic of the first singlet excited state of PtPOP [23]. This leads us to think that, at the particular conditions at which the experiment was performed, all contribution to the observed dynamics from Pt–Pt coherent vibrations in the excited state is suppressed, while the oscillatory behaviour in the difference signal must arise from motion in the ground-state potential surface. In Chap. 10, we will examine in depth how the choice of the off-resonant pump pulse shown in Fig. 7.2 created the conditions to probe excusively ground-state structural dynamics upon photoexcitation. In Chap. 12, we will show

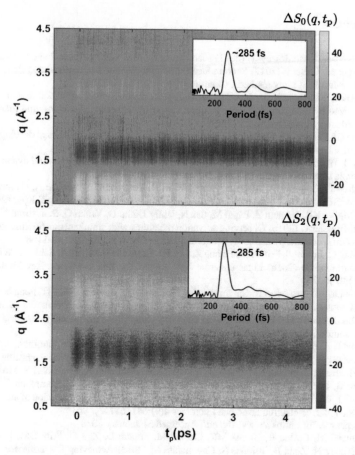

Fig. 7.3 Isotropic ($\Delta S_0(q, t_p)$, top) and anisotropic ($\Delta S_2(q, t_p)$, bottom) difference scattering signals as obtained from ultrafast XDS measurements upon photoexcitation of PtPOP to the S_1 state in water. The signals are expressed in electronic units (e.u.) per solute molecule. Electronic units represent the scattering of a free electron, i.e. the Thomson cross-section. The insets show the FT of the first right-singular vector of an SVD analysis of each of the two sets of difference scattering signals

that a picture of nonequilibrium dynamics produced using QM/MM BOMD trajectories in the S_0 and S_1 states of PtPOP in water substantiates the hypothesis we have just formulated.

References

1. Als-Nielsen J, McMorrow D (2011) Elements of modern x-ray physics, 2nd edn. Wiley
2. Chergui M, Collet E (2017) Photoinduced structural dynamics of molecular systems mapped by time-resolved x-ray methods. Chem Rev 117(16):11025–11065
3. Kim KH, Kim J, Lee JH, Ihee H (2014) Topical review: molecular reaction and solvation visualized by time-resolved x-ray solution scattering: structure, dynamics, and their solvent dependence. Struct Dynam 1:011301
4. Milne CJ, Penfold TJ, Chergui M (2014) Recent experimental and theoretical developments in time-resolved x-ray spectroscopies. Coordination Chem Rev 277:44–68
5. Ihee H, Wulff M, Kim J, Adachi S (2010) Ultrafast x-ray scattering: structural dynamics from diatomic to protein molecules. Int Rev Phys Chem 29:453–520
6. Haldrup K, Gawelda W, Abela R, Alonso-Mori R, Bergmann U, Bordage A, Cammarata M, Canton SE, Dohn AO, Van Driel TB, Fritz DM, Galler A, Glatzel P, Harlang T, Kjær KS, Lemke HT, Møller KB, Németh Z, Pápai M, Sas N, Uhlig J, Zhu D, Vankó G, Sundström V, Nielsen MM, Bressler C (2016) Observing solvation dynamics with simultaneous femtosecond x-ray emission spectroscopy and x-ray scattering. J Phys Chem B 120(6):1158–1168
7. Bostedt C, Boutet S, Fritz DM, Huang Z, Lee HJ, Lemke HT, Robert A, Schlotter WF, Turner JJ, Williams GJ (2016) Linac coherent light source: the first five years. Rev Modern Phys 88(1):015007
8. Chollet M, Alonso-Mori R, Cammarata M, Damiani D, Defever J, Delor JT, Feng Y, Glownia JM, Langton JB, Nelson S, Ramsey K, Robert A, Sikorski M, Song S, Stefanescu D, Srinivasan V, Zhu D, Lemke HT, Fritz DM (2015) The x-ray pump-probe instrument at the linac coherent light source. J Synchrotron Radiat 22:503–507
9. Emma P, Akre R, Arthur J, Bionta R, Bostedt C, Bozek J, Brachmann A, Bucksbaum P, Coffee R, Decker FJ, Ding Y, Dowell D, Edstrom S, Fisher A, Frisch J, Gilevich S, Hastings J, Hays G, Ph Hering Z, Huang R, Iverson H, Loos M, Messerschmidt A, Miahnahri S, Moeller HD, Nuhn G, Pile D, Ratner J, Rzepiela D, Schultz T, Smith P, Stefan H, Tompkins J, Turner J, Welch J, White W, Wu J, Yocky G, Galayda J (2010) First lasing and operation of an ångstrom-wavelength free-electron laser. Nat Photonics 4(9):641–647
10. European XFEL. https://www.xfel.eu/. Accessed 31 January 2018
11. Harmand M, Coffee R, Bionta MR, Chollet M, French D, Zhu D, Fritz DM, Lemke HT, Medvedev N, Ziaja B, Toleikis S, Cammarata M (2013) Achieving few-femtosecond time-sorting at hard x-ray free-electron lasers. Nat Photonics 7(3):215–218
12. Møller KB, Henriksen NE (2012) Time-resolved x-ray diffraction: the dynamics of the chemical bond. Struct Bonding 142:185–212
13. Lakowicz JR (2006) Principles of fluorescence spectroscopy, 3rd edn. Springer, New York
14. Lorenz U, Møller KB, Henriksen NE (2010) On the interpretation of time-resolved anisotropic diffraction patterns. New J Phys 12:113022
15. Baskin JS, Zewail AH (2006) Oriented ensembles in ultrafast electron diffraction. Chem Phys Chem 7(7):1562–1574
16. Baskin JS, Zewail AH (2005) In space and time. Chem Phys Chem 6:226–2276
17. Van Kleef EH, Powis I (1999) Anisotropy in the preparation of symmetric top excited states. I. One-photon electric dipole excitation. Mol Phys 96(5):757–774
18. Biasin E, van Driel TB, Levi G, Laursen MG, Dohn AO, Moltke A, Vester P, Hansen FBK, Kjaer KS, Hartsock R, Christensen M, Gaffney KJ, Henriksen NE, Møller KB, Haldrup K, Nielsen MM (2018) Anisotropy enhanced x-ray scattering from solvated transition metal complexes. J Synchrotron Radiat 25(2):306–315
19. Biasin E, van Driel TB, Kjær KS, Dohn AO, Christensen M, Harlang T, Chabera P, Liu Y, Uhlig J, Pápai M, Németh Z, Hartsock R, Liang W, Zhang J, Alonso-Mori R, Chollet M, Glownia JM, Nelson S, Sokaras D, Assefa TA, Britz A, Galler A, Gawelda W, Bressler C, Gaffney KJ, Lemke HT, Møller KB, Nielsen MM, Sundström V, Vankó G, Wärnmark K, Canton SE, Haldrup K (2016) Femtosecond x-ray scattering study of ultrafast photoinduced structural dynamics in solvated [Co(terpy)2]2+. Phys Rev Lett 117(1):013002

20. Haldrup K, Levi G, Biasin E, Vester P, Laursen MG, Beyer F, Kjær KS, Brandt van Driel T, Harlang T, Dohn AO, Hartsock RJ, Nelson S, Glownia JM, Lemke HT, Christensen M, Gaffney KJ, Henriksen NE, Møller KB, Nielsen MM (2019) Ultrafast x-ray scattering measurements of coherent structural dynamics on the ground-state potential energy surface of a diplatinum molecule. Phys Rev Lett 122:063001

21. van Driel TB, Kjær KS, Biasin E, Haldrup K, Lemke HT, Nielsen MM (2015) Disentangling detector data in XFEL studies of temporally resolved solution state chemistry. Faraday Discuss 177:443

22. Che CM, Butler LG, Gray HB, Crooks RM, Woodruff WH (1983) Metal-metal interactions in binuclear platinum(II) diphosphite complexes. Resonance Raman spectra of the $1A_{1g}(d\sigma^*)^2$ and $3A_{2u}(d\sigma^*p\sigma)$ electronic states of $(Pt_2(P_2O_5H_2)_4^{4-})$. J Am Chem Soc 105(16):5492–5494

23. van der Veen RM, Cannizzo A, van Mourik F, Vlček A Jr, Chergui M (2011) Vibrational relaxation and intersystem crossing of binuclear metal complexes in solution. J Am Chem Soc 113:305

Chapter 8
Simulating and Analysing X-Ray Diffuse Scattering Signals

Due to the lack of long-range periodic order in solution, it is not possible to reconstruct the structure of a liquid sample directly from an X-ray scattering pattern, as one would do in X-ray crystallography [1–3]. Instead, one has to rely on a suitable structural model to fit the experimental data. In XDS pump-probe measurements the information on the structural changes upon photoirradiation are condensed in 1D difference scattering curves, which are inherently dominated by scattering from distributions of interatomic distances. This makes the determination of the structural changes undergone by solvated molecules particularly challenging.

In this chapter, we will illustrate how atomistic modelling brings aid to the analysis and interpretation of ultrafast XDS data. In order to do so, we show how it is possible to calculate time-dependent scattering signals from molecular structures and nuclear distributions generated by BOMD simulations. We will focus, in particular, on the modelling strategy employed to analyse the XDS data collected in the pump-probe XFEL experiments on PtPOP in aqueous solution, described in Sect. 7.2 of Chap. 7. Before doing that, in the following section, we shall briefly mention the main results of the theory of time-resolved X-ray scattering of molecules. The details of processing and analysing X-ray solution scattering data can be found in Refs. [1–4, 6–8]. The theoretical framework for interpretation of time-resolved X-ray scattering experiments has been formulated over the past two decades. Key works in this context are represented by Refs. [9–14].

8.1 Calculating X-Ray Scattering Signals

Ignoring all inelastic contributions and quantum coherence effects, and further neglecting the frequency spread of the X-ray probe pulse (known as the "static approximation"), the time-dependent X-ray scattering signal of a system of N_n atoms at a pump-probe time delay t_p can be expressed as [11]:

© Springer Nature Switzerland AG 2019
G. Levi, *Photoinduced Molecular Dynamics in Solution*, Springer Theses,
https://doi.org/10.1007/978-3-030-28611-8_8

$$S(\mathbf{q}, t_\mathrm{p}) = \int \tilde{\varrho}(\mathbf{R}, t_\mathrm{p}) \mid F(\mathbf{R}, \mathbf{q}) \mid^2 \mathrm{d}\mathbf{R} \tag{8.1}$$

where $S(\mathbf{q}, t_\mathrm{p})$ is in units of the Thomson cross-section of a free electron, and $\tilde{\varrho}(\mathbf{R}, t_\mathrm{p})$ is given by the instantaneous probability distribution of nuclear geometries created by the pump pulse averaged over the intensity profile $I(t - t_\mathrm{p})$ of the X-ray pulse:

$$\tilde{\varrho}(\mathbf{R}, t_\mathrm{p}) = \int_0^\infty I(t - t_\mathrm{p})\varrho(\mathbf{R}, t)\mathrm{d}t \tag{8.2}$$

Equation (8.1) involves also the molecular form factor $F(\mathbf{R}, \mathbf{q})$, which represents the scattering from a static nuclear configuration and is related to a Fourier transform of the electron density of the system [11]. Under the assumption that the electron density is unchanged upon excitation, Eq. (8.1) is valid for a system of molecules distributed over different electronic states ($\varrho(\mathbf{R}, t) = \sum_n \varrho_n(\mathbf{R}, t)$, where $\varrho_n(\mathbf{R}, t)$ is the nuclear probability distribution associated with state n). This approximation is satisfied within the Independent Atom Model (IAM). According to the IAM, the electron density is the sum of spherical atomic densities, and $F(\mathbf{R}, \mathbf{q})$ is given by:

$$F(\mathbf{R}, \mathbf{q}) = \sum_a^{N_n} f_a(q)e^{i\mathbf{q}\mathbf{R}_a} \tag{8.3}$$

where f_a is an atomic form factor for atom a. Despite neglecting chemical bonding between atoms, the IAM provides reasonable results in most cases [15]. In an X-ray scattering calculation, the atomic form factors are taken from tabulated values [16]. In what follows, we focus the discussion on the case of excitation to only one excited state.

8.1.1 Difference Scattering Curves from BOMD Distributions

In Sect. 7.1 of the previous chapter, we have said that the difference scattering signal is composed of a term due to scattering from the excited-state ensemble of molecules and a term that is the signature of the hole left in the ground state by the laser (see Eq. (7.6)). From Eq. (8.1) we can obtain Eq. (7.6) using nuclear distributions:

$$\begin{aligned}
\Delta S(\mathbf{q}, t_\mathrm{p}) &= S_\mathrm{on}(\mathbf{q}, t_\mathrm{p}) - S_\mathrm{off}(\mathbf{q}, t_\mathrm{p}) \\
&= \int \left[\tilde{\varrho}_\mathrm{on}(\mathbf{R}, t_\mathrm{p}) - \tilde{\varrho}_\mathrm{off}(\mathbf{R}, t_\mathrm{p})\right] \mid F(\mathbf{R}, \mathbf{q}) \mid^2 \mathrm{d}\mathbf{R} \\
&= \int \left\{\alpha\tilde{\varrho}_\mathrm{ES}(\mathbf{R}, t_\mathrm{p}) + \left[\tilde{\varrho}_\mathrm{GS}^\mathrm{eq}(\mathbf{R}) - \alpha\tilde{\varrho}_\mathrm{GS}^\mathrm{h}(\mathbf{R}, t_\mathrm{p})\right] - \tilde{\varrho}_\mathrm{GS}^\mathrm{eq}(\mathbf{R})\right\} \mid F(\mathbf{R}, \mathbf{q}) \mid^2 \mathrm{d}\mathbf{R} \\
&= \int \alpha \left[\tilde{\varrho}_\mathrm{ES}(\mathbf{R}, t_\mathrm{p}) - \tilde{\varrho}_\mathrm{GS}^\mathrm{h}(\mathbf{R}, t_\mathrm{p})\right] \mid F(\mathbf{R}, \mathbf{q}) \mid^2 \mathrm{d}\mathbf{R} \\
&= \alpha \left[S_\mathrm{ES}(\mathbf{q}, t_\mathrm{p}) - S_\mathrm{GS}^\mathrm{h}(\mathbf{q}, t_\mathrm{p})\right]
\end{aligned} \tag{8.4}$$

where we have used that the nuclear probability distribution created by the pump pulse is given by the distribution in the excited state plus the difference between the ground-state equilibrium distribution and the ground-state hole.

In order to simulate the difference scattering signal of a photoexcited molecular ensemble we need to calculate $S_{ES}(\mathbf{q}, t_p)$ and $S_{GS}^h(\mathbf{q}, t_p)$. Furthermore, to account for the polarization of the pump pulse, we need to consider the decomposition into an isotropic and an anisotropic terms, according to Eq. (7.7). It can be shown [4, 12], using Eq. (8.1) and the IAM (Eq. (8.3)), that the isotropic scattering signal of a species s, where for species we indicate either the excited state or the ground-state hole, can be computed from:

$$S_0^s(q, t_p) = \sum_{a,b}^{N_n} f_a(q) f_b(q) 4\pi \int_0^\infty d^2 \tilde{\varrho}_{ab}^s(d, t_p) \frac{\sin(qd)}{qd} d(d) \qquad (8.5)$$

where $\tilde{\varrho}_{ab}^s(d, t_p)$ is the time-averaged pairwise probability distribution function of the distance d between atoms a and b, and $f_a(q)$ and $f_b(q)$ are their respective form factors. Note that in Eq. (8.5) there is no vectorial dependence, in accordance with the fact that the isotropic contribution of the scattering signal from the decomposition Eq. (7.7) is equivalent to the scattering of a randomly oriented ensemble of molecules [12].

Equation (8.5) can be recast in a more computationally convenient form, in which the sums run over all atom types (where an atom type correspond to a particular element) and the pairwise distribution functions are replaced by radial distribution functions (RDFs) [4]. RDFs are readily obtained from BOMD simulations. The relation that is used for the conversion is the definition of RDF between atom types l and m as the ratio between a probability distribution function $\tilde{\varrho}_{lm}^s(d, t_p)$ that collects the probability distributions $\tilde{\varrho}_{ab}^s(d, t_p)$, where atoms a and b belong to atom types l and m, respectively, and the isotropic density ϱ_{lm}^0:

$$\tilde{g}_{lm}^s(d, t_p) = \frac{\tilde{\varrho}_{lm}^s(d, t_p)}{\varrho_{lm}^0} \qquad (8.6)$$

where ϱ_{ab}^0 is the inverse of the volume V of the BOMD simulation box ($\varrho_{ab}^0 = 1/V$). The isotropic scattering signal in terms of RDFs is given by [4]:

$$S_0^s(q, t_p) = \sum_l N_n^l f_l^2(q) + \sum_{l,m} f_l(q) f_m(q) \frac{N_n^l (N_n^m - \delta_{lm})}{V} 4\pi$$

$$\times \int_0^{R_{box}} d^2 \left[\tilde{g}_{lm}^s(d, t_p) - g_{lm}^0 \right] \frac{\sin(qd)}{qd} d(d) \qquad (8.7)$$

in which N_n^l and N_n^m are the total number of atoms of type l and m, respectively, R_{box} is the length of the simulation box, and g_{lm}^0 is the homogeneous density limit of the RDF. All details of the above derivation can be found in Ref. [4]. Equation (8.7) has been implemented in the Matlab programming language by former PhD student and Postdoc A. O. Dohn, who has co-supervised the present project in our group [4]. For practical applications, when one is interested in a difference scattering signal, it is more convenient to use directly the following relation:

$$\Delta S_0(q, t_p) = \alpha \sum_{l,m} f_l(q) f_m(q) \frac{N_n^l \left(N_n^m - \delta_{lm} \right)}{V} 4\pi$$

$$\times \int_0^{R_{box}} d^2 \Delta \tilde{g}_{lm}(d, t_p) \frac{\sin(qd)}{qd} \mathrm{d}(d) \tag{8.8}$$

where $\Delta \tilde{g}_{lm}(d, t_p)$ is the difference RDF given by:

$$\Delta \tilde{g}_{lm}(d, t_p) = \tilde{g}_{lm}^{ES}(d, t_p) - \tilde{g}_{lm}^{GS,h}(d, t_p) \tag{8.9}$$

We have, therefore, modified the original script by A. O. Dohn to compute difference scattering signals according to Eq. (8.8).

The anisotropic contribution to the scattering signal can be computed from a formula that involves angle-dependent pairwise probability distribution functions ($\tilde{\varrho}_{ab}^s(d, \vartheta, t_p)$, where ϑ is the angle between the interatomic distance vector $\mathbf{d}$ and the transition dipole moment of the molecule), as shown in Ref. [12]:

$$S_2^s(q, t_p) = - c_2^s(t_p) \sum_{a,b}^{N_n} f_a(q) f_b(q) 2\pi$$

$$\times \int_0^\infty \int_0^\pi d^2 \sin(\vartheta) \tilde{\varrho}_{ab}^s(d, \vartheta, t_p) P_2(\cos \vartheta) j_2(qd) \mathrm{d}(d) \mathrm{d}(\vartheta) \tag{8.10}$$

where j_2 is the second spherical Bessel function ($j_2(x) = \left(\frac{3}{x^2} - 1 \right) \frac{\sin(x)}{x} - \frac{3\cos(x)}{x^2}$), and the coefficient $c_2^s(t_p)$ describes the rotational distribution of the species s. The time dependence of $c_2^s(t_p)$ is dictated by the rotational correlation time of s [17]. We have not implemented an equivalent formula using angle-resolved RDFs from BOMD simulations, yet. We note, on the other hand, that to model the anisotropic part of the difference scattering data of PtPOP we found adequate to only employ single molecular structures, in which case $\tilde{\varrho}_{ab}^s(d, \vartheta, t_p)$ in Eq. (8.10) is a delta function, as it will be explained in the next section.

8.2   Analysis of Ultrafast XDS Data

The strategy employed by our experimental collaborators for the quantitative analysis of measured 1D difference XDS curves, like those shown in Fig. (7.2), is to compare the signal to simulated curves within a maximum-likelihood framework [1, 2]. In other words, to infer the structural changes one minimises the difference between the experimental data and the simulated signal by optimizing the structural model. The model includes three terms:

$$\Delta S^{\text{model}}(q, t_p) = \Delta S^{\text{solu}}(q, t_p) + \Delta S^{\text{solu}-\text{solv}}(q, t_p) + \Delta S^{\text{solv}}(q, t_p) \qquad (8.11)$$

The first term, $\Delta S^{\text{solu}}(q, t_p)$, models changes in the distances between atoms of the solute. Due to the very low concentrations at which XDS experiments are performed, one neglects the scattering arising from interferences between different solute molecules and considers only intramolecular distances [3]. The second term, $\Delta S^{\text{solu}-\text{solv}}(q, t_p)$, is the scattering due to changes in solute-solvent distances. It reflects rearrangements in the solvation shell around the solute, and it is, for this reason, often referred to as cage term. The third contribution to the model, $\Delta S^{\text{solv}}(q, t_p)$, accounts for changes in distances between atoms of the solvent, which usually reflect changes in the thermodynamic state of the bulk solvent.

We now show how the isotropic and anisotropic contributions to each of the terms in Eq. (8.11) are typically computed, and highlight the particular choices made in the analysis of the PtPOP XDS data. In the calculation of the various terms we neglect the finite duration of the X-ray probe pulse in Eq. (8.2), thus assuming that time averaged nuclear distributions are equal to the instantaneous distributions.

8.2.1 The Solute Term

We first examine how the isotropic contribution to the scattering is modelled. $\Delta S_0^{\text{solu}}(q, t_p)$ is given by the isotropic scattering of the excited state minus the isotropic scattering from the ground-state hole:

$$\Delta S_0^{\text{solu}}(q, t_p) = \alpha \left[S_{\text{ES},0}^{\text{solu}}(q, t_p) - S_{\text{GS},0}^{h,\text{solu}}(q, t_p) \right] \qquad (8.12)$$

Here, the assumption is usually made that the two terms are the scattering of classical single structures. This amounts to disregarding entirely the spread of the nuclear distribution functions involving atoms of the solute. For a pair of atoms a and b the pairwise probability distribution function becomes:

$$4\pi d^2 \tilde{\varrho}_{ab}^s(d, t_p)\text{d}(d) = \delta(d - d_{ab}^s(t_p))\text{d}(d) \qquad (8.13)$$

When Eq. (8.13) is inserted into Eq. (8.5), we obtain the formula that describes $S_{0,\text{ES}}^{\text{solu}}(q, t_p)$ and $S_{0,\text{GS}}^{h,\text{solu}}(q, t_p)$ under the approximation of neglecting the spread of the distributions:

$$S_{s,0}^{\text{solu}}(q, t_p) = \sum_{a,b}^{N_n^{\text{solu}}} f_a(q) f_b(q) \frac{\sin[q d_{ab}^s(t_p)]}{q d_{ab}^s(t_p)} \qquad (8.14)$$

which is the well-known Debye formula for the orientationally averaged scattering of a gas-phase molecule [18]. We will see how this approximation is justified in the case of PtPOP in Chap. 11, when we will compare the signal calculated from single gas-phase structures with the scattering obtained from RDFs generated through QM/MM BOMD simulations. For the fitting of the experimental data, the scattering signal is calculated for sets of ground- and excited-state structures generated by varying a selection of structural parameters in the molecule. The risk of overfitting the data limits the number of explored parameters to only a few, usually those involving the atoms that scatters the most in the molecule, and that are expected to undergo significant changes upon excitation.

In the case of the analysis of the PtPOP XDS data, the model for the solute assumed a fixed excited-state structure and incorporated the time dependence through the Pt-Pt distance (d_{PtPt}) of a (delta-function) ground-state hole:

$$\Delta S_0^{solu}(q, t_p) = \alpha \left[S_{ES,0}^{solu}(q) - S_{GS,0}^{h,solu}(q, d_{PtPt}(t_p)) \right] \tag{8.15}$$

We will see more in detail how the structure for the excited state and those for the ground state were constructed in Chap. 11. The excitation fraction α, which is a constant for the time scales considered in the analysis, can be treated as a free parameter in the fitting. However, α and structural parameters in the fit are known to be strongly correlated [2, 19]. For this reason, α was first determined by analysing the difference scattering signal at a pump-probe time delay of 4.5 ps, at which both ground and excited states are known to have reached vibrational equilibrium. Afterwards, α was locked and the structural fitting of the time-dependent signal employed the Pt-Pt distance of the ground-state hole as the only free parameter.

The anisotropic contribution to the difference scattering signal of the solute is given by:

$$\Delta S_2^{solu}(q, t_p) = \alpha \left[S_{ES,2}^{solu}(q, t_p) - S_{GS,2}^{h,solu}(q, t_p) \right] \tag{8.16}$$

Also in this case, the signals from the excited and ground states are computed using single structure. Equation (8.10) with delta functions instead of nuclear distribution functions becomes:

$$S_{s,2}^{solu}(q, t_p) = -c_2(t_p) \sum_{a,b}^{N_n^{solu}} f_a(q) f_b(q) P_2[\cos \vartheta_{ab}(t_p)] j_2[q d_{ab}(t_p)] \tag{8.17}$$

The model that was used to fit the measured time-dependent anisotropic difference scattering signal of PtPOP has the same form of Eq. (8.15):

$$\Delta S_2^{solu}(Q, t) = \alpha \left[S_{ES,2}^{solu}(q) - S_{GS,2}^{h,solu}(q, d_{PtPt}(t_p)) \right] \tag{8.18}$$

Once again, the structural dynamics is parametrized only through the Pt–Pt distance of the (delta-function) hole representing depletion of the ground-state ensemble.

8.2.2 The Solute-Solvent Term

Due to the larger number of degrees of freedom (DOF) involved, the signal from solute-solvent interferences cannot be described using single structures, and one has to appeal to BOMD simulations to generate nuclear distributions. The strategy that we commonly employ consists in determining the difference signal only once using equilibrium ground- and excited-state distributions obtained from equilibrated BOMD data; while the dynamics is modelled as a fraction of the final (equilibrium) value through a scaling factor $\beta(t)$. For the isotropic signal:

$$\Delta S_0^{\text{solu-solv}}(q, t_p) = \beta(t)\Delta S_{\text{MD}}^{\text{solu-solv}}(q) \qquad (8.19)$$

where $\beta(t)$ grows from zero to the excitation fraction α. $\Delta S_{\text{MD}}^{\text{solu-solv}}(q)$ in Eq. (8.19) is computed from Eq. (8.8) with $\Delta \tilde{g}_{lm}(d, t_p)$ (see Eq. (8.9)) obtained from excited-state equilibrium RDFs ($\tilde{g}_{lm}^{\text{ES}}(d, t_p) = g_{lm}^{\text{ES}}(d, t) = g_{lm}^{\text{ES,eq}}(d)$) and ground-state equilibrium RDFs ($\tilde{g}_{lm}^{\text{GS,h}}(d, t_p) = g_{lm}^{\text{GS,h}}(d, t) = g_{lm}^{\text{GS,eq}}(d)$, where we assume that the hole distributions have the same form of the equilibrium ground-state distributions). In the computation of $\Delta \tilde{g}_{lm}(d, t_p)$, the index l for the first summation in Eq. (8.8) runs over atom types within the solute and the index m for the second summation runs over atom types of the solvent.

For the analysis of PtPOP we observed that there was no need to include the solute-solvent term in the fit of $\Delta S_2(q, t_p)$, since the quality of the fit was already sufficiently good without [17]. Therefore, this term was only taken into account in the modelling of the isotropic contribution to the difference scattering signal through Eq. (8.19). The RDFs to compute $\Delta S_{\text{MD}}^{\text{solu-solv}}(q)$ were obtained from equilibrium ground- and excited-state QM/MM BOMD data, as we will see in Chap. 11.

8.2.3 The Solvent Term

The solvent term arises from changes in the thermodynamic variables of the bulk solvent (temperature (T), density (ρ) or pressure (p)). Variations of these parameters happen as an effect of energy transfer from the solute to the solvent during the excited-state relaxation events [7], and are isotropic. Therefore, one has to account for such changes only in the isotropic part of the difference scattering signal. This is done by determining in separate reference measurements the differential of the scattering with respect to two of the three thermodynamic variables [7]. Usually one measures the differential of the pressure and of the temperature.

Since changes in pressure of the solvent are known to take place on nanosecond time scales [3], well beyond the range of pump-probe delays explored in the PtPOP experiments, in the modelling of the isotropic solvent difference scattering signal we included only a term due to changes in temperature at constant pressure:

$$\Delta S_0^{solv}(q, t_p) = \Delta T(t_p) \frac{\partial \Delta S_0^{ref}(q)}{\partial T}\bigg|_\rho \tag{8.20}$$

8.2.4 Summarising the Model for the PtPOP Data

Collecting all terms presented in the previous paragraphs, the model used to analyse the entire set of time-resolved XDS data collected in the XFEL experiments on PtPOP in water described in Sect. 7.2 is:

$$\Delta S_0^{model}(q, t_p) = \alpha \left[S_{0,ES}^{solu}(q) - S_{0,GS}^{h,solu}(q, d_{PtPt}(t_p)) \right]$$

$$+ \beta(t) \Delta S_{MD}^{solu-solv}(q) + \Delta T(t_p) \frac{\partial \Delta S_0^{ref}(q)}{\partial T}\bigg|_\rho \tag{8.21}$$

$$\Delta S_2^{model}(q, t_p) = \alpha \left[S_{2,ES}^{solu}(q) - S_{2,GS}^{h,solu}(q, d_{PtPt}(t_p)) \right] \tag{8.22}$$

Finally, we note that the general procedure outlined here to analyse 1D XDS curves does not guarantee to reach a true global minimum in the optimization of the model with respect to the experimental data. Prior experimental knowledge and the support of QM/MM BOMD simulations are fundamental in this regard. For example, the choice of modelling a time-dependent ground-state hole in Eqs. (8.21) and (8.22), while fixing the structure of the excited state for PtPOP, was motivated by a comparison of the period of the oscillating signal with the known Pt-Pt vibrational period for the ground and excited states of the molecule, and by the guidance offered by QM/MM BOMD simulations, which could confirm the hypothesis that the observed dynamics is the signature of a moving ground-state hole. The assistance of QM/MM BOMD simulations is even more important for the determination of structural changes involving the solvation shell, and hence modelling of the term $\Delta S^{solu-solv}(q, t_p)$, for which prior knowledge is often lacking. In Chap. 11 we will see more in detail how calculation of $\Delta S^{solu-solv}(q, t_p)$ from QM/MM BOMD simulations made up for deficiencies in the model of the $\Delta S_0(q, t_p)$ signal of PtPOP.

References

1. Haldrup K, Nielsen MM (2014) Measuring and understanding ultrafast phenomena using X-rays, pp 91–113. NATO Science for Peace and Security Series A: Chemistry and Biology. Springer, Netherlands

2. Haldrup K, Christensen M, Nielsen MM (2010) Analysis of time-resolved x-ray scattering data from solution-state systems. Acta Crystallogr A 66:261–369

3. Ihee H, Wulff M, Kim J, Adachi S (2010) Ultrafast x-ray scattering: structural dynamics from diatomic to protein molecules. Int Rev Phys Chem 29:453–520

4. Dohn AO, Biasin E, Haldrup K, Nielsen MM, Henriksen NE, Møller KB (2015) On the calculation of x-ray scattering signals from pairwise radial distribution functions. J Phys B Atomic Mol Opt Phys 48(24):244010. Corrigendum: [5]

5. Dohn AO, Biasin E, Haldrup K, Nielsen MM, Henriksen NE, Møller KB (2016) Corrigendum: On the calculation of x-ray scattering signals from pairwise radial distribution functions. J Phys B Atomic Mol Opt Phys 48(24):059501

6. Levantino M, Yorke BA, Monteiro DCF, Cammarata M, Pearson AR (2015) Using synchrotrons and XFELs for time-resolved X-ray crystallography and solution scattering experiments on biomolecules. Curr Opin Struct Biol 35:41–48

7. Kjaer KS, van Driel TB, Kehres J, Haldrup K, Khakhulin D, Bechgaard K, Cammarata M, Wulff M, Sorensen TJ, Nielsen MM (2013) Introducing a standard method for experimental determination of the solvent response in laser pump, X-ray probe time-resolved wide-angle X-ray scattering experiments on systems in solution. Phys Chem Chem Phys 15:15003–15016

8. Lee JH, Kim KH, Kim TK, Lee Y, Ihee H (2006) Analyzing solution-phase time-resolved x-ray diffraction data by isolated-solute models. J Chem Phys 125:174504

9. Simmermacher M, Henriksen NE, Møller KB (2017) Time-resolved X-ray scattering by electronic wave packets: analytic solutions to the hydrogen atom. Phys Chem Chem Phys 19(30):19740–19749

10. Kirrander A, Saita K, Shalashilin DV (2016) Ultrafast X-ray scattering from molecules. J Chem Theor Comput 12(3):957–967

11. Møller KB, Henriksen NE (2012) Time-resolved x-ray diffraction: the dynamics of the chemical bond. Struct Bonding 142:185–212

12. Lorenz U, Møller KB, Henriksen NE (2010) On the interpretation of time-resolved anisotropic diffraction patterns. New J Phys 12:113022

13. Henriksen NE, Møller KB (2008) On the theory of time-resolved x-ray diffraction. J Phys Chem B 112:558–567

14. Ben-Nun M, Cao J, Wilson KR (1997) Ultrafast x-ray and electron diffraction: theoretical considerations. J Phys Chem A 101(47):8743–8761

15. Coppens P (1992) Electron density from x-ray diffraction. Ann Rev Phys Chem 43:663–692

16. Cromer DT, Mann JB (1968) X-ray scattering factors computed from numerical Hartree-Fock wave functions. Acta Crystallogr A 24:321–324

17. Biasin E, van Driel TB, Levi G, Laursen MG, Dohn AO, Moltke A, Vester P, Hansen FBK, Kjaer KS, Hartsock R, Christensen M, Gaffney KJ, Henriksen NE, Møller KB, Haldrup K, Nielsen MM (2018) Anisotropy enhanced x-ray scattering from solvated transition metal complexes. J Synchrotron Radiat 25(2):306–315

18. Als-Nielsen J, McMorrow D (2011) Elements of modern x-ray physics, 2nd edn. Wiley

19. Haldrup K, Levi G, Biasin E, Vester P, Laursen MG, Beyer F, Kjær KS, Brandt van Driel T, Harlang T, Dohn AO, Hartsock RJ, Nelson S, Glownia JM, Lemke HT, Christensen M, Gaffney KJ, Henriksen NE, Møller KB, Nielsen MM (2019) Ultrafast x-ray scattering measurements of coherent structural dynamics on the ground-state potential energy surface of a diplatinum molecule. Phys Rev Lett 122:063001

Part IV
Simulations Results

Chapter 9
Gas-Phase Molecular Geometry

9.1 Computational Details of the GPAW Calculations

In all PtPOP simulations that are reported in this thesis, unless otherwise specified, the electronic structure of the complex was calculated using GPAW [2, 3] with representation of the KS molecular orbitals in a basis of linear combination of atomic orbitals (LCAO) [4].

The excited states were described with the ΔSCF scheme presented in Sect. 5.5. We chose a σ of 0.01 eV for the Gaussian smearing of the orbital occupation numbers. In the tests of the ΔSCF implementation (Sect. 5.5), this value of σ was found to bring no detectable changes in the PES of the lowest-lying singlet state of the CO molecule for geometries at which ΔSCF without Gaussian smeared constraints could converge (see Fig. 5.8). A σ of 0.01 eV allowed to readily converge all steps of all ΔSCF trajectories of PtPOP. For the open-shell singlet excited state, the calculations employed the spin-unpolarized approach described in Sect. 5.5. Spin-unpolarized calculations are computationally much cheaper for geometry optimizations and BOMD simulations than Ziegler's sum method [5], because the latter requires SCF convergence of two single-determinant states, one having mixed singlet-triplet and one with triplet spin symmetry.

The exchange-correlation functional employed in the calculations was the GGA functional BLYP [6, 7], while the basis functions were tzp [4] for the Pt atoms and dzp [4] for all other atoms. We used a grid spacing of the GPAW cell of 0.18 Å. This choice of LCAO basis set and grid spacing ensures that the structure of the complex is converged with respect to these simulation parameters, as shown in the next section.

Parts of this chapter have been reproduced with permission from Ref. [1], https://doi.org/10.1021/acs.jpcc.8b00301. Copyright 2018 American Chemical Society.

9.2   Preliminary Studies

9.2.1   GPAW Convergence Tests

Electronic structure calculations are usually preceded by preparatory tests. When using GPAW, we must ensure that (i) the size of the simulation cell is large enough for a sufficiently accurate representation of the orbitals, and (ii) the molecular properties of interest are converged with respect to spacing between points of the real-space grid and, in the case of LCAO calculations, quality of the basis set.

The size of the cell containing the real-space grid on which numerical basis functions and electron density of PtPOP were represented, was chosen such that the distance between any of the atoms of the complex and any of the cell borders was at least 6 Å. This choice of cell size was seen to be sufficient to eliminate spurious effects due to truncation of the KS orbitals at the borders.

Since we used LCAO calculations in all PtPOP simulations, we tested convergence with respect to both the grid spacing and atomic orbital basis set. Convergence was tested against the main structural parameter of interest, namely the Pt–Pt distance, as shown in Fig. 9.1. Deviations smaller than 1% with respect to the Pt–Pt distance

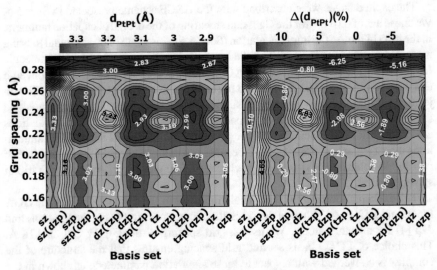

Fig. 9.1 Convergence of the Pt–Pt distance of PtPOP in the ground state with respect to grid spacing and LCAO basis set size. For each combination of grid spacing/basis set the geometry of the molecule was fully optimized in vacuum with a convergence criteria of 0.02 eV/Å for the maximum force on all individual atoms. The syntax "sz(dzp)" indicates that the calculations employed single-zeta (sz) functions from a double-zeta polarized (dzp) basis set. (Left) Contour plot of the Pt–Pt distance. (Right) Contour plot of the percentage deviation of the Pt–Pt distance with respect to the value obtained when using a 0.15 Å grid spacing with qzp basis set. A deviation of 1% corresponds to an absolute difference of 0.03 Å. Satisfactory convergence is achieved for grid spacings smaller than 0.20 Å and only when using polarized basis

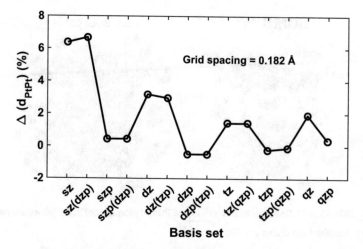

Fig. 9.2 Convergence of the Pt–Pt distance with respect to the size of the basis set. Each point is obtained from a geometry optimization in vacuum at a grid spacing of 0.182 Å with different size of the basis set. The y values express the percentage deviation of the Pt–Pt distance with respect to the Pt–Pt distance obtained when using a 0.150 Å grid spacing with qzp basis set. Therefore, the plot corresponds to a horizontal cut at 0.182 Å of the contour plot in Fig. 9.1(Right). This value of grid spacing is that utilized in all GPAW calculations performed on PtPOP in the present work

obtained with the smallest grid spacing (0.15 Å) and largest basis set (qzp), are achieved when employing grid spacings smaller than 0.20 Å and basis sets larger than single-zeta and including polarization functions. The importance of including polarization functions for an accurate description of the structure of the complex is highlighted in Fig. 9.2.

Given the above results, the GPAW simulations of PtPOP could safely employ a 0.18 Å grid spacing and basis set of tzp quality for the Pt atoms and dzp for all other atoms.

9.2.2 Molecular Orbitals and Electron Density

Figure 9.3 shows a depiction of the HOMO and LUMO orbitals of the ground-state PtPOP molecule fully optimized in vacuum with GPAW. The shape of the orbitals clearly reflects their metal-metal $d\sigma^*$ and $p\sigma$ character, respectively: the HOMO is antibonding in the region between the two Pt atoms and mainly localized outwards along the Pt-Pt axis, the LUMO is σ-bonding, while also extending on the outer sides of the PtP$_4$ faces, a feature that is attributable to the involvement of p orbitals of the phosphorus atoms [8].

Promotion of an electron from the HOMO to the LUMO leads to formation of the S$_1$ and T$_1$ excited states of the molecule. Figure 9.4 illustrates the effects of excitation to the S$_1$ state on the electron density of the complex. The $d\sigma^* \rightarrow p\sigma$

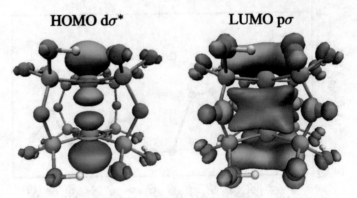

Fig. 9.3 HOMO and LUMO molecular orbitals of PtPOP ground state at the gas-phase optimized geometry. Isovalues are drawn at $0.075 \sqrt{e^-/\text{Å}^3}$

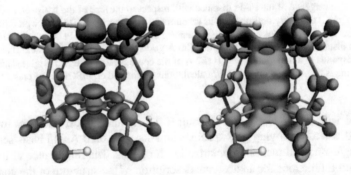

Fig. 9.4 Difference between the electron density of the S_1 and S_0 states of PtPOP at the ground-state GPAW optimized geometry of the complex (electron density S_1—electron density S_0). Isovalues are drawn at $0.0056\,e^-/\text{Å}^3$. (Left) Negative part of the difference density. (Right) Positive values. The integral over the volume of the positive or negative part of the difference density gives a value of $\pm 0.56\,e^-$

transition results in build-up of electron density between the two Pt atoms and loss of it along the Pt–Pt axis in outward position. Besides, gain of electron density in S_1 is also apparent close to the P atoms, along the outer sides of the PtP_4 faces, reflecting the involvement of p ligand orbitals in the formation of the LUMO. This outward shift of density compensates in part for the loss along the Pt–Pt axis.

9.2.3   Excitation Energies

Table 9.1 reports computed vacuum $S_0 \rightarrow S_1$ and $S_0 \rightarrow T_1$ vertical excitation energies together with the respective experimental values.

Table 9.1 Vertical $S_0 \to S_1$ and $S_0 \to T_1$ excitation energies of PtPOP in vacuum calculated using ΔSCF in GPAW and comparison with experimental values retrieved from the literature. The calculations were performed at the optimized ground-state vacuum geometry of the complex. All values are in eV

	Calc	Exp [9, 10]
$S_0 \to S_1$	3.50[a] 3.51[b]	3.35–3.44
$S_0 \to T_1$	3.27	2.72–2.76
$\Delta(S_1 - T_1)$	0.23[a] 0.24[b]	0.63–0.68

[a]Computed using Ziegler's sum rule [5]
[b]Obtained from spin-unpolarized calculations

The transition energies were calculated with ΔSCF in GPAW at the S_0 gas-phase optimized structure of PtPOP. For the S_1 state we used both the spin-unpolarized technique, which is the method we employed in all BOMD simulations of PtPOP, and Ziegler's sum rule.

Calculated vertical $S_0 \to S_1$ transition energies are within $\sim$3% of the experimental range of values obtained from the maximum of the $S_0 \to S_1$ band of absorption spectra of crystals [9, 10]. Almost exact agreement is found between the S_1 excitation energies computed using the two different ΔSCF methods to describe the singlet excited state. This confirms that the spin-unpolarized ΔSCF description of the S_1 state of PtPOP has the same level of accuracy as calculations based on the more computationally expensive sum rule.

The calculated T_1 excitation energy is $\sim$20% larger than the experimental values. This, in turn, causes the S_1–T_1 splitting to be underestimated by a factor of around 2.5. A similar excitation energy for the triplet was obtained by Novozhilova et al. [11] by TDDFT with the BLYP functional and an all-electron basis set for Pt. We note that in the present study we focus on the dynamics in the S_1 state happening at times shorter than the known intersystem crossing (ISC) time of PtPOP in water ($\sim$14 ps [12]). All excited-state BOMD simulations did not account for singlet-triplet transitions induced by spin orbit couplings (SOCs) and, therefore, they were not affected by the underestimation of the S_1–T_1 energy gap.

9.3   Ground- and Excited-State Geometries

The geometries of the ground state (S_0) and lowest-lying singlet (S_1) and triplet (T_1) excited states of PtPOP were fully optimized in vacuum using a quasi-Newton local optimization algorithm implemented in ASE. A previous DFT study [13] identified two conformers of PtPOP in the ground state with staggered (C_{4h} symmetry) and eclipsed (D_4 symmetry) hydrogen bonding motifs, respectively, the eclipsed structure being about 0.036 eV more stable at the DFT-B3LYP level. In the present work, we optimized the more stable ground state conformation; this structure was then used as a starting point to optimize the geometry in the excited states. Geometry optimization

Fig. 9.5 Visualization of the
PtPOP complex with the
atomic labels used to
indicate the structural
parameters reported in
Table 9.2. The molecular
structure represented here
corresponds to the geometry
fully optimized in the ground
state with GPAW

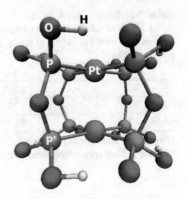

was carried out until the maximum force on all individual atoms was less than 0.02
eV/Å. Fully-optimized geometries were confirmed to be true minima of the potential
energy surface by inspection of the frequencies of a normal mode analysis.

The S_0 and T_1 geometries optimized in ASE and GPAW were compared to optimized geometries obtained using a standard implementation of KS DFT within the
Gaussian09 program package [14]. The calculations in Gaussian09 were performed
by Postdoc Mátyás Pápai in our group. In these calculations, the unrestricted formalism was used to describe the excited state. We employed the Ahlrichs TZVP [15]
all-electron basis set for the P, O, H atoms, and the quasirelativistic effective core
potential (ECP) def2-ECP [16] in conjunction with the valence electrons Ahlrichs
def2-TZVP [17] basis set for the Pt atoms. Two different exchange-correlation functionals were used: the BLYP functional, which was also utilized in the GPAW calculations, and the commonly employed hybrid functional B3LYP [18, 19], to test the
effect of including a portion of exact Hartree-Fock exchange energy on the structure
of the complex. Also for these calculations, we checked the frequencies of a normal
mode analysis to confirm that the fully-optimized geometries were true minima.

Potential energy curves in vacuum, in a particular electronic state, were computed
by scanning along relevant coordinates, starting from the fully-optimized geometry
of that state, while relaxing at each step all other degrees of freedom with the same
convergence criteria as used in the full geometry optimizations in ASE.

9.3.1   Optimized Structures

Relevant structural parameters of the S_0 state of PtPOP (see Fig. 9.5 for a depiction
of the molecule) together with the differences with respect to the S_1 and T_1 structures
obtained from the geometry optimizations in vacuum, are given in Table 9.2.

The ground state is found to have approximate C_{4h} symmetry, with a D_{4h} Pt_2P_8
core. The largest discrepancy between the S_0 structure predicted using GPAW and
the one obtained using more conventional atom-centered basis sets and an ECP for

Table 9.2 Selected structural parameters for the S_0, S_1 and T_1 states of PtPOP obtained from the geometries optimized in vacuum at different DFT levels

| | S_0 | | | | $\Delta (T_1 - S_0)$[b] | | | $\Delta (S_1 - S_0)$ |
| | ECP/TZVP[a] | | | | ECP/TZVP[a] | | | |
	BLYP GPAW	BLYP	B3LYP	BLYP GPAW	BLYP	B3LYP		BLYP GPAW
Bond (Å)								
Pt–Pt	3.005	3.091	3.065	−0.211	−0.241	−0.248		−0.205
Pt–P	2.393	2.425	2.399	0.031	0.032	0.019		0.032
P–O(–P′)	1.718	1.711	1.679	0.001	0.002	−0.000		0.001
P...P′	3.098	3.126	3.084	−0.060	−0.062	−0.067		−0.055
Angles (deg)								
P–O–P′	128.84	131.90	133.38	−4.63	−5.07	−5.39		−4.27
(Pt–Pt–P)$_\alpha$	91.14	90.40	90.23	5.49	5.45	4.42		5.42
(Pt–Pt–P)$_\beta$	91.08	90.42	90.22	−1.96	−1.28	−0.14		−1.93
P–Pt–Pt–P′	0.03	0.00	0.01	0.61	0.54	0.36		0.46

[a]ECP and valence electrons basis set used for the Pt atoms
[b]T_1 calculated using unrestricted DFT

Pt with the same exchange-correlation functional, is in the Pt–Pt distance, which is 0.086 Å shorter in the GPAW structure. We note that the GPAW calculated Pt–Pt distance of 3.005 Å is around 0.09 Å closer to the midpoint of the experimental range (2.913-2.979 Å) of values found from X-ray crystallography [20–23]. The differences become smaller in the excited states since the structure calculated with standard KS DFT experiences a larger Pt–Pt contraction.

As already mentioned before in this thesis, the Pt–Pt contraction in the excited states is a consequence of excitation of an electron from the metal-metal HOMO anti-bonding to the metal-metal LUMO bonding orbital. Eventually, the Pt–Pt contractions from ground to excited state predicted by all different methods are well within the experimental range (0.19–0.28 Å) of values obtained from Franck-Condon analysis of the vibronic progression of low-temperature absorption and emission spectra [10, 24], and X-ray diffraction measurements of crystals [20, 25]. The Pt–Pt bond in the T_1 state is found to be shorter than in the S_1 state of ∼0.01 Å from the GPAW calculations. Indeed, a slightly reduced contraction in the singlet excited state with respect to the triplet has been inferred experimentally by comparing the wavenumbers of the Pt–Pt stretching progression exhibited by the absorption bands of crystal (n-Bu$_4$N)$_4$[PtPOP] relative to the S_1 (145–147 cm^{-1}) and T_1 (150 cm^{-1}) states [8, 9], and was further confirmed by the DFT calculations performed by Zális et al. [26] using the PBE0 functional, which delivered a $\Delta(S_1 - T_1)$ for the Pt–Pt bond of ∼0.02 Å.

Turning to the other geometrical parameters, the most prominent changes between ground- and excited-state structures in interatomic distances involving atoms of the ligands are represented by a lengthening of the Pt–P bonds and by a shortening of

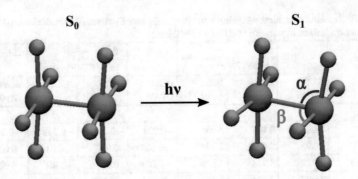

Fig. 9.6   The Pt_2P_8 core of PtPOP of the geometries of the S_0 and S_1 states fully optimized in vacuum with GPAW. In the ground state, each PtP_4 group is in a local square pyramidal geometry and the core has D_{4h} symmetry. Following excitation to S_1, the core distorts towards a D_{2d} conformation, in which PtP_4 moieties are in a quasi-trigonal bipyramidal geometry

the $P\cdots P'$ distances along the Pt–Pt axis, which is, however, much smaller than the Pt–Pt contraction itself. An elongation of the Pt–P bonds in the excited state of PtPOP is a well-known prediction of DFT [11]. Since using B3LYP results in a ~40% smaller elongation, as evident from Table 9.2, it seems also to be the structural effect of excitation that is most sensible to the introduction of exact exchange in the DFT functional. Apart from that, BLYP and B3LYP predicted structural changes from ground to excited state agree within 0.007 Å for bond lengths and 1° for angles, while differences in the ground state are all smaller than 2% of the BLYP calculated values. Therefore, given the similarities between BLYP and B3LYP results in this case, it was possible to perform the BOMD simulations using the computationally cheaper GGA functional without considerable loss of accuracy with respect to DFT with a hybrid functional.

An interesting aspect of the optimized geometry of the excited states that emerges from inspection of the values of the ∠Pt–Pt–P angles reported in Table 9.2 is that PtP_4 moieties do not retain a local square-based planar geometry but slightly distort towards a quasi-trigonal bipyramidal structure. As a consequence, in the excited states the symmetry of the Pt_2P_8 core is lowered to D_{2d}. This is underpinned by the fact that ∠Pt–Pt–P angles in the excited states do not have the same value, as in the ground state, where they are approximately 90°. Instead, one can define ∠Pt–Pt–P angles involving equatorial and axial P atoms in a local quasi-trigonal bipyramidal geometry. The distortion is represented in Fig. 9.6, where we have indicated ∠Pt–Pt–P angles involving equatorial and axial P atoms as α and β, respectively. We will discuss in more detail this symmetry lowering involving the ligands of the complex in the next section, where the distortion will be characterized by means of PES scans.

9.3.2 Potential Energy Surfaces

Figure 9.7 shows the PESs computed along the Pt–Pt coordinate for all three electronic states using GPAW with the BLYP functional. As expected, the PESs of T_1 and S_1 are shifted to shorter Pt–Pt distances with respect to the ground state and parallel to each other. A feature that, up to now, had only been postulated experimentally based on the similarities between the low-temperature $S_0 \rightarrow S_1$ and $S_0 \rightarrow T_1$ absorption bands [8, 9]. To our knowledge, this is the first time that this experimental observation is confirmed by a DFT calculation of the S_1 and T_1 PESs of PtPOP. Given the close similarities between the electronic structures of d^8–d^8 complexes [8], we can argue that also the other members of this class of compounds feature parallel T_1 and S_1 PESs. In light of this, the choice of using gradients calculated in the first triplet state to mimic BO dynamics in S_1, as previously done in our group to simulate by unrestricted DFT the S_1 dynamics of a diiridium d^8–d^8 complex [27], appears justified, at least for simulations in vacuum.

The relative energies of the singlet and triplet excited states are affected by the understabilization of the triplet by ~ 0.5 eV that has already been noted in Sect. 9.2. However, since we performed only BOMD simulations in the S_1 state for times considerably shorter than the ISC times observed for PtPOP in water solution, reproducing an accurate energy picture of the lowest triplet excited state was not relevant for these studies.

We have already mentioned that, in the excited states, PtP_4 groups of the molecule arrange according to a local quasi-trigonal bipyramidal geometry. The extent of the distortion can be quantified by the difference (indicated by Δ) between $\angle$Pt–Pt–P angles involving equatorial and axial P atoms of the local quasi-trigonal bipyrami-

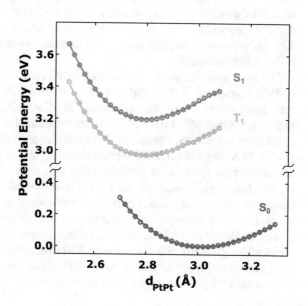

Fig. 9.7 PESs along the Pt–Pt coordinate computed in vacuum for the S_0, S_1 and T_1 states of PtPOP using GPAW. Open circles represent the calculated points, while the lines are 3rd order polynomial fits

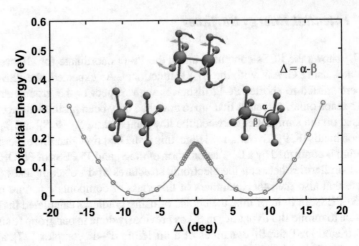

Fig. 9.8 Vacuum PES of the S_1 state of PtPOP along the pseudorotation coordinate Δ. Δ is the angle difference defined in the figure. The Pt_2P_8 core of PtPOP is shown at the symmetric minima and at the transition state of the potential energy curve. The structure at $\Delta = 0$ has D_{4h} symmetry and each PtP$_4$ group is in a local square pyramidal geometry, as in the fully optimized ground-state molecule. The minima correspond to a Pt_2P_8 core with D_{2d} symmetry and PtP$_4$ groups in a quasi-trigonal bipyramidal geometry. Open circles represent the calculated points, while the line is a cubic spline fit to the data

dal geometry (see Fig. 9.8). To characterize in more detail this structural distortion involving the ligands, we have computed the PES in the S_1 state along the coordinate Δ. The PES is shown in Fig. 9.8 and clearly reveals the presence of a rotational barrier between equivalent D_{2d} geometries. The pseudorotation of the P atoms in each PtP$_4$ group resembles the Berry isomerization mechanism [28] occurring in trigonal bipyramidal molecules, although the angle $\angle$(Pt–Pt–P)$_\alpha$ does not reach the 120° value characteristic of a perfect bipyramidal geometry due to the rigidity of the P-O-P bridging ligands.

We have investigated the dependence of this prediction of DFT on the choice of DFT functional. To this end, we have calculated the PES of PtPOP in the S_1 state along the pseudorotation coordinate Δ using ΔSCF in GPAW with the PBE functional, which differs from BLYP in that it does not include empirically optimized parameters as the latter. The results of the calculation are presented in Fig. 9.9, where also the PES computed at BLYP level, and already shown in Fig. 9.8, is included. The two functionals agree in the prediction that structures with PtP$_4$ groups in a quasi-trigonal bipyramidal geometry are located at energy-minima, although the height of the barrier for pseudorotation between the two equivalent isomers predicted by PBE is observed to be almost 4 times smaller than that computed with BLYP.

D_{2d} isomers of transition metal M_2L_8 dimers, where each ML$_4$ is in a local trigonal bipyramidal geometry and can undergo Berry pseudorotation, are known [29], but have never been reported before for PtPOP. From the experimental side, Ohashi and co-workers [20] interpreted the outcome of time-resolved X-ray diffraction

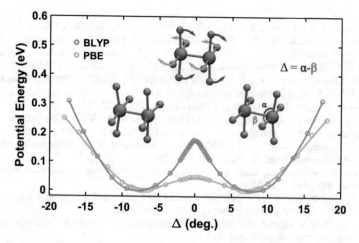

Fig. 9.9 PESs of PtPOP in the S_1 state along the angle difference Δ defined in figure. The PESs were computed in vacuum using ΔSCF in GPAW with the BLYP and PBE functionals by relaxing at each step all degrees of freedom apart from Δ. Open circles are the calculated points, while the lines are cubic spline fits to the data

measurements of crystals assuming D_{4h} symmetry. However, the analysis derived a large contraction of ~0.1–0.2 Å of the Pt–P bonds, which is in contrast to the slight lengthening obtained from all DFT calculations. Moreover, it should be considered that in crystals there are packing forces and interactions with counterions that might come into play, which are not taken into account in the calculations of the gas-phase isolated molecule, making the validity of a direct comparison with experiments dubious. It is difficult, on the other hand, to explain why previous computational works where the structure of PtPOP in the triplet state was optimized with unrestricted DFT without symmetry constraints, have not reported this ligand distortion with symmetry lowering of the Pt_2P_8 core. The existence of a local minimum at a geometry with D_{2d} symmetry for both the T_1 and S_1 states is supported by all type of DFT calculations presented here, and was reproduced also by a GGA functional different than BLYP. In the absence of detailed information about the true nature of reported T_1 geometries in the literature, we speculate that this ligand distortion might have been overlooked.

References

1. Levi G, Pápai M, Henriksen NE, Dohn AO, Møller KB (2018) Solution structure and ultrafast vibrational relaxation of the PtPOP complex revealed by ΔSCF-QM/MM direct dynamics simulations. J Phys Chem C 122:7100–7119
2. Mortensen JJ, Hansen LB, Jacobsen KW (2005) Real-space grid implementation of the projector augmented wave method. Phys Rev B 71:035109

3. Enkovaara J, Rostgaard C, Mortensen JJ, Chen J, Dulak M, Ferrighi L, Gavnholt J, Glinsvad C, Haikola V, Hansen HA, Kristoffersen HH, Kuisma M, Larsen AH, Lehtovaara L, Ljungberg M, Lopez-Acevedo O, Moses PG, Ojanen J, Olsen T, Petzold V, Romero NA, Stausholm-Møller J, Strange M, Tritsaris GA, Vanin M, Walter M, Hammer B, Häkkinen H, Madsen GKH, Nieminen RM, Nørskov JK, Puska M, Rantala TT, Schiøtz J, Thygesen KS, Jacobsen KW (2010) Electronic structure calculations with GPAW: a real-space implementation of the projector augmented-wave method. J Phys Condens Matter 22:253202

4. Larsen AH, Vanin M, Mortensen JJ, Thygesen KS, Jacobsen KW (2009) Localized atomic basis set in the projector augmented wave method. Phys Rev B 80:195112

5. Ziegler T, Rauk A, Baerends EJ (1977) On the calculation of multiplet energies by the hartree-fock-slater method. Theor Chimica Acta 43(3):261–271

6. Becke AD (1988) Density-functional exchange-energy approximation with correct asymptotic behavior. Phys Rev A 38:3098

7. Lee C, Yang W, Parr RG (1988) Development of the Colle-Salvetti correlation-energy formula into a functional of the electron density. Phys Rev B 37:785–789

8. Gray HB, Záliš S, Vlček A (2017) Electronic structures and photophysics of d8–d8 complexes. Coordination Chem Rev 345:297–317

9. Stiegman AE, Rice SF, Gray HB, Miskowski VM (1987) Electronic spectroscopy of d^8–d^8 diplatinum complexes. $^1a_{2u}(d\sigma^* \rightarrow p\sigma)$, $^3e_u(d_{xz},d_{yz} \rightarrow p\sigma)$, and $^{3,1}b_{2u}(d\sigma^* \rightarrow d_{x^2-y^2})$ excited states of $pt_2(p_2o_5h_2)_4^{4-}$. Inorg Chem 26:1112

10. Rice SF, Gray HB (1983) Electronic absorption and emission spectra of binuclear platinum(II) complexes. Characterization of the lowest singlet and triplet excited states of $Pt_2(P_2O_5H_2)_4^{4-}$. J Am Chem Soc 105:4571–4575

11. Novozhilova IV, Volkov AV, Coppens P (2003) Theoretical analysis of the triplet excited state of the [Pt2(H2P2O5)4]4- ion and comparison with time-resolved X-ray and spectroscopic results. J Am Chem Soc 125(4):1079–1087

12. van der Veen RM, Cannizzo A, van Mourik F, Vlček A Jr, Chergui M (2011) Vibrational relaxation and intersystem crossing of binuclear metal complexes in solution. J Am Chem Soc 113:305

13. Gellene GI, Roundhill DM (2002) Computational studies on the isomeric structures in the pyrophosphito bridged diplatinum (II) complex, platinum pop. J Phys Chem A 106:7617–7620

14. G09—Gaussian 09, Revision D.01, Frisch MJ, Trucks GW, Schlegel HB, Scuseria GE, Robb MA, Cheeseman JR, Scalmani G, Barone V, Petersson GA, Nakatsuji H, Li X, Caricato M, Marenich A, Bloino J, Janesko BG, Gomperts R, Mennucci B, Hratchian HP, Ortiz JV, Izmaylov AF, Sonnenberg JL, Williams-Young D, Ding F, Lipparini F, Egidi F, Goings J, Peng B, Petrone A, Henderson T, Ranasinghe D, Zakrzewski VG, Gao J, Rega N, Zheng G, Liang W, Hada M, Ehara M, Toyota K, Fukuda R, Hasegawa J, Ishida M, Nakajima T, Honda Y, Kitao O, Nakai H, Vreven T, Throssell K, Montgomery Jr JA, Peralta JE, Ogliaro F, Bearpark M, Heyd JJ, Brothers E, Kudin KN, Staroverov VN, Keith T, Kobayashi R, Normand J, Raghavachari K, Rendell A, Burant JC, Iyengar SS, Tomasi J, Cossi M, Millam JM, Klene M, Adamo C, Cammi R, Ochterski JW, Martin RL, Morokuma K, Farkas O, Foresman JB, Fox DJ (2016) Gaussian, Inc., Wallingford CT

15. Schäfer A, Horn R, Ahlrichs R (1992) Fully optimized contracted gaussian basis sets for atoms Li to Kr. J Chem Phys 97:2571–2577

16. Andrae D, Häußermann U, Dolg M, Stoll H, Preuß H (1990) Energy-adjusted ab initio pseudopotentials for the second and third row transition elements. Theor Chim Acta 77(2):123–141

17. Weigend F, Ahlrichs R (2005) Balanced basis sets of split valence, triple zeta valence and quadruple zeta valence quality for H to Rn: design and assessment of accuracy. Phys Chem Chem Phys 7:3297–3305

18. Becke AD (1993) Density-functional thermochemistry. III. The role of exact exchange. J Chem Phys 98:5648–5652

19. Stephens PJ, Devlin FJ, Chabalowski CF, Frisch MJ (1994) Ab-initio calculation of vibrational absorption and circular-dichroism spectra using density-functional force-fields. J Phys Chem 98:11623–11627

20. Yasuda N, Uekusa H, Ohashi Y (2004) X-ray analysis of excited-state structures of the diplatinum complex anions in five crystals with different cations. Bull Chem Soc Jpn 77(5):933–944

21. Ozawa Y, Terashima M, Mitsumi M, Toriumi K, Yasuda N, Uekusa H, Ohashi Y (2003) Photoexcited crystallography of diplatinum complex by multiple-exposure IP method. Chem Lett 32(1):62–63

22. Che CM, Herbstein FH, Schaefer WP, Marsh RE, Gray HB (1983) Binuclear platinum diphosphite complexes. Crystal structures of K4[Pt2(pop)4Br]·3H2O, a new linear chain semiconductor, and K4[Pt2(pop)4Cl2]·2H2O. J Am Chem Soc 105(14):4604–4607

23. Sadler J, Sanderson MR (1980) A novel di-platinum(II) octaphosphite complex showing metal-metal bonding and intense luminescence; a potential probe for basic proteins. X-ray crystal and molecular structure. J Chem Soc Chem Commun (4):13–15

24. Ikeyama T, Yamamoto S, Azumi T (1988) Vibrational analysis of sublevel phosphorescence spectra of potassium tetrakis(μ-diphosphonato)diplatinate(II): mechanism of radiative transition for the electronically forbidden A1u spectrum. J Phys Chem 92(24):6899

25. Kim CD, Pillet S, Wu G, Fullagar WK, Coppens P (2002) Excited-state structure by time-resolved X-ray diffraction. Acta Crystallogr Sect A Found Crystallogr 58:133–137

26. Záliš S, Lam Y-C, Gray HB, Vlček A (2015) Spin orbit TDDFT electronic structure of diplatinum(II, II) complexes. Inorg Chem 54:3491–3500

27. Dohn AO, Jónsson EÖ, Kjær KS, van Driel TB, Nielsen MM, Jacobsen KW, Henriksen NE, Møller KB (2014) Direct dynamics studies of a binuclear metal complex in solution: the interplay between vibrational relaxation, coherence, and solvent effects. J Phys Chem Lett 5:2414–2418

28. Berry RS (1960) Correlation of rates of intramolecular tunneling processes, with application to some group V compounds. J Chem Phys 32(3):933–938

29. Alain D, Albright TA, Hoffmann R (1979) Some hydrido-bridged transition-metal dimers and their unsupported analogues. Speculations on pentuple bonding and pentuple bridging. J Am Chem Soc 101(12):3141–3151

Chapter 10
Computational Details of the QM/MM BOMD Simulations

10.1 Equilibrium Ground-State Simulations

The optimized geometry of PtPOP in its ground electronic state (S_0) was placed in a cubic simulation box with side length of 35 Å containing TIP4P [1] water molecules at a density of 1 g/cm^3 pre-equilibrated in the NVT ensemble at 300 K. The total number of solvent molecules, after removing those overlapping with the solute, was 1383. The QM subsystem was defined to comprise only the complex. The MM subsystem included the TIP4P water molecules plus four K$^+$ counterions to neutralize the total charge of the box. Potassium was also used as counterion in the time-resolved X-ray diffuse scattering (XDS) experiments on PtPOP in water presented in Chap. 7. The counterions were described as point charges, using force field parameters from Ref. [2]. During the dynamics, the position of each counterion was restrained to regions of the simulation box outside a sphere centered at the center of the QM cell by applying the restraint potential shown in Eq. (6.31). The cutoff radius d_{pr} and the harmonic force constant k_{pr} for the restraint potential were 16 Å and 500 kcal/mol respectively. This choice of d_{pr} and k_{pr} ensured that the atoms of the complex were at least 12 Å apart from the counterions during the simulations, as seen from the solute-K$^+$ radial distribution functions (RDFs). The ground state of PtPOP was described using the BLYP functional, a grid spacing of 0.18 Å, and with tzp basis set [3] for Pt and dzp [3] for the rest of the atoms. Non-bonded dispersion and exchange repulsion interactions between the solute and the MM particles were parametrized through the standard Lennard-Jones (LJ) potential of Eq. (6.5), using for the atoms of the complex LJ parameters from the universal force field (UFF) [4].

After solvating the complex, the entire box was further equilibrated in NVT to 300 K, employing a 1 fs time step until stabilization of the temperature. Thermalization was realized using the Langevin thermostat implemented in ASE. The thermostat was applied only to the solvent, while the friction on the atoms of the solute was set to 0. Periodic boundary conditions were applied according to the minimum image convention. Stability of the simulations was ensured by constraining all OH bonds and hydrogen bonds present in PtPOP with the RATTLE algorithm [5]. After the equilibration, QM/MM BOMD data were collected with 2 fs time step in the NVT ensemble with the thermostat applied to the solvent, for at least further 25 ps.

© Springer Nature Switzerland AG 2019
G. Levi, *Photoinduced Molecular Dynamics in Solution*, Springer Theses,
https://doi.org/10.1007/978-3-030-28611-8_10

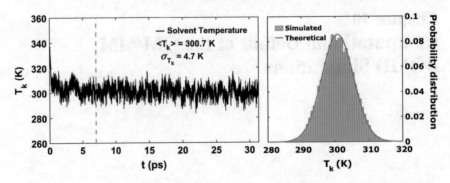

Fig. 10.1 (Left) Equilibration of the instantaneous kinetic temperature of the solvent for a QM/MM BOMD trajectory of PtPOP in the ground state. The vertical grey dashed line represents the time at which the trajectory was considered equilibrated. The average instantaneous kinetic temperature $\langle T_k \rangle$ and the variance σ_{T_k} were computed over the equilibrated part of the trajectory. (Right) The distribution of instantaneous kinetic temperatures from the equilibrated trajectory as compared to a Gaussian probability distribution with mean $\langle T_k \rangle$ and variance calculated according to Eq. (10.3)

Figure 10.1 (Left) shows the equilibration of the instantaneous kinetic temperature of the solvent in the course of the simulation. The instantaneous kinetic temperature T_k at a time t during a trajectory propagation of a collection of N_{MM} atoms can be calculated from:

$$T_k(t) = \frac{1}{(3N_{MM} - N_c)\,k_b} \sum_{k=1}^{N_{MM}} M_k \mid \dot{\mathbf{R}}_k(t) \mid^2 \tag{10.1}$$

where N_c is the total number of constrained internal degrees of freedom (DOF). For N_{H_2O} TIP4P water molecules $N_{MM} = N_c = 3N_{H_2O}$, since in the TIP4P force field all three internal DOF of H_2O are constrained. Therefore, Eq. (10.1) expressed as a function of the total number of water molecules in the simulation box becomes:

$$T_k(t) = \frac{1}{6N_{H_2O}k_b} \sum_{k=1}^{N_{MM}} M_k \mid \dot{\mathbf{R}}_k(t) \mid^2 \tag{10.2}$$

The average instantaneous kinetic temperature $\langle T_k \rangle$ over the equilibrated part of the trajectory, which is taken at times $t > 7$ ps, was equal to 300.7 K. Figure 10.1 (Right) compares the distribution of instantaneous kinetic temperatures from the equilibrated part of the trajectory with the theoretical probability distribution for a canonical ensemble at $\langle T_k \rangle$. The theoretical distribution was obtained as a Gaussian function with variance calculated from $\langle T_k \rangle$ according to Boltzmann statistics for the NVT ensemble [6]:

$$\sigma_{NVT}^2(T_k) = \frac{2\langle T_k \rangle^2}{3N_{MM} - N_c} = \frac{\langle T_k \rangle^2}{3N_{H_2O}} \tag{10.3}$$

The perfect agreement between the simulated and theoretical distributions indicates that the simulations are able to reproduce the correct fluctuations of the instantaneous kinetic temperature for an NVT ensemble.

From the equilibrated part of the trajectory, 48 more parallel QM/MM BOMD production runs were started at 0.5 ps intervals, to further accelerate the data collection process. When starting each trajectory, the velocities of the atoms in the solvent were randomized by imposing a Maxwell-Boltzmann distribution at 300 K, to minimize the correlation between them. Overall, the equilibrated trajectories amounted to 460 ps and were obtained over ~9750 h of CPU time, corresponding to ~21 h per picosecond.

To assess the impact of constraining all OH and hydrogen bonds in the complex on equilibrium properties and dynamics, a single trajectory with increased mass for all hydrogen atoms but no constraints on the degrees of freedom of the solute was produced. The average of the main structural parameters of the complex and the Pt–Pt oscillating frequency obtained from this trajectory were found to be negligibly different from those obtained when employing RATTLE constraints for the OH and hydrogen bonds.

10.2 Nonequilibrium Dynamics Due to Laser Excitation

The structural and solvation dynamics following photoexcitation by an ultrashort pulse of PtPOP in water is investigated using classical trajectories of the nuclei obtained by computing energy and forces at QM/MM level. The basic principle we base our study on is that individual trajectories or ensembles of them generated or selected from already available equilibrium BOMD ensembles, according to out-of-equilibrium initial conditions, reflect the laser-induced dynamics of the system.

The ground-state simulations detailed in the previous section established a large set of around 230000 equilibrated QM/MM BOMD snapshots collected over a total simulation time of about 460 ps. Collectively these data add up to an equilibrium ensemble of ground-state PtPOP configurations in water, from which initial conditions for the nonequilibrium dynamics in the ground and S_1 excited states can be drawn.

Excitation to the S_1 state by an ultrashort optical pulse was described in a picture of instantaneous promotion of ground-state molecules from the underlying equilibrium ground-state distribution of Pt–Pt distances ($P_{GS}^{eq}(d_{PtPt})$) to S_1 according to a spatial filtering (SF) approximation [7–11] of the pump-pulse transition. This approximation takes into account the frequency distribution of the ultrashort pulse but neglects any effect due to nuclear motion throughout its finite temporal duration. For a Gaussian pulse $\epsilon(t) \propto e^{-\frac{t^2}{2\tau^2}} e^{-i\omega_1 t}$, where ω_1 and τ are respectively the center frequency and temporal width, initial conditions for a set of excited-state trajectories can be sampled from the (unnormalized) distribution $P_{ES}(d_{PtPt}, t_0)$ given by:

$$P_{ES}(d_{PtPt}, t_0) = F^2(d_{PtPt}) P_{GS}^{eq}(d_{PtPt}) \tag{10.4}$$

In Eq. (10.4), the excitation window $F(d_{PtPt})$ takes the form:

$$F(d_{PtPt}) = A \exp \left[-\frac{\tau^2 (\Delta V(d_{PtPt}) - \hbar\omega_1)^2}{2\hbar^2} \right] \tag{10.5}$$

where $\Delta V(d_{PtPt})$ is the potential energy difference between the ground and excited states. An ultrashort pump pulse burns a hole in the ground-state equilibrium distribution of Pt–Pt distances. The change in the ground-state distribution $P_{GS}^{eq}(d_{PtPt})$ induced by the pulse, which we will call distribution of the hole, will exhibit periodic motion in the ground-state PES with the characteristic period of the ground state [11]. Following Fleming [11], we approximate the distribution of the hole at time zero $P_{GS}^{h}(d_{PtPt}, t_0)$ with $P_{ES}(d_{PtPt}, t_0)$, i.e. we assume that the hole left in ground-state distribution by the pulse has the same form of the non-stationary distribution created in the excited state.

The procedure that we used to generate initial conditions for the nonequilibrium dynamics in the ground and excited states within the SF approximation is the following:

1. Construct $\Delta V(d_{PtPt})$. The choice of the potentials is not trivial, since the concept itself of fixed potential energy surfaces is ambiguous in the context of BOMD simulations in the presence of a solvent (we will examine the assumption of using fixed potential energy surfaces in the next paragraph). A reasonable choice is to use the free energy surface of PtPOP along the Pt–Pt coordinate obtained as potential of mean force (PMF) from equilibrium QM/MM BOMD simulations. While for the ground state the PMF is readily available from the set of equilibrated ground-state trajectories, for the excited state the PMF is not known before performing excited-state QM/MM BOMD simulations. Establishing a set of equilibrated excited-state trajectories before the nonequilibrium propagation would be computationally expensive. So, we computed $\Delta V(d_{PtPt})$ from parameters known from steady-state and ultrafast optical measurements in solution.
2. Choose the parameters ω_1 and τ of the optical pump pulse.
3. Obtain $P_{ES}(d_{PtPt}, t_0) = P_{GS}^{h}(d_{PtPt}, t_0)$ from Eqs. (10.4) and (10.5).
4. Propagate ΔSCF-QM/MM BOMD trajectories in the excited state starting from an ensemble of ground-state equilibrium configurations reflecting $P_{ES}(d_{PtPt}, t_0)$. In the selection of ground-state frames we ensured that they were spaced at least 0.5 ps from each other, such to minimize the correlation between excited-state trajectories.
5. Remove an ensemble of PtPOP molecules reflecting $P_{GS}^{h}(d_{PtPt,0})$ from the ground-state equilibrium ensemble. The remaining ground-state molecules represent a non-stationary ensemble.

We established two sets of nonequilibrium ground- and excited-state ensembles using two different choices of laser parameters. For the first set of initial conditions, the laser pump pulse was chosen to reproduce as close as possible the experimental conditions of the pump-probe X-ray diffuse scattering (XDS) measurements of PtPOP in water presented in Sect. 7.2, in which ground-state dynamics was observed. The initial conditions for the second set of excited-state nonequilibrium simulations and the corresponding nonequilibrium ensemble of remaining ground-state molecules employed the parameters of the laser used by van der Veen et al. [12] in femtosecond transient absorption measurements of the ultrafast excited-state vibrational dynamics of PtPOP in water. In what follows, we examine the details of the initial conditions of each set of ensembles, and at the end compare them.

The first set of initial conditions was aimed at modelling ground-state hole dynamics of PtPOP in water for helping the analysis and substantiating the outcome of the ultrafast XDS measurements we performed at the LCLS XFEL of Stanford [13, 14]. $\Delta V(d_{PtPt})$ was taken as the difference between two harmonic potentials with force constants calculated from the reduced mass of the Pt2 dimer and the vibrational frequencies obtained by van der Veen et al. [12] using femtosecond transient absorption measurements in water solution, which in wavenumbers are 119 and 149 cm^{-1} for S_0 and S_1, respectively. For the position of the minima of the potentials, the Pt–Pt distances of the S_0 and S_1 gas-phase optimized geometries were used (see Table 9.2); finally, the two potentials were shifted relative to each other such that the energy difference at the Pt–Pt distance of the optimized ground-state geometry was equal to 3.35 eV (corresponding to a wavelength of ~370 nm), i.e. the transition energy at the maximum of the $S_0 \rightarrow S_1$ band of the experimental absorption spectrum in aqueous solution [13, 15, 16]. The parameters for the excitation field were obtained from a Gaussian fit to the spectral intensity profile of the pump pulse that was used in the XDS experiment. Under the assumption that the pulse is Fourier-transform limited, the fit gave a τ of 20 fs; while for $\hbar\omega_l$ we obtained 3.14 eV (~395 nm). The Gaussian fit to the experimental pump pulse is shown in Fig. 10.2 together with the experimental absorption spectrum of PtPOP in water [13]. In order to remove a fraction of ground-state molecules reflecting the experimental excitation fraction α, the parameter A in the expression of the excitation window, Eq. (10.5), can be increased until the desired value of α is obtained. Adopting this procedure can lead to complete depopulation of the ground-state equilibrium distribution at specific Pt–Pt distances. Since the experiment employed a linearly polarized excitation pulse, we took into account the fact that the orientation dependence of the absorption probability [17] limits the number of molecules that can be excited. Thus, in order to avoid the unphysical situation of depopulating entirely the ground state at a particular Pt–Pt distance, the S_1 distribution given by Eq. (10.4) was modified according to:

$$P'_{ES}(d_{PtPt}, t_0) = \begin{cases} \frac{1}{B} P_{ES}(d_{PtPt}, t_0) & \text{if } P_{ES}(d_{PtPt}, t_0) > \frac{1}{B} P^{eq}_{GS}(d_{PtPt}) \\ P_{ES}(d_{PtPt}, t_0) & \text{if } P_{ES}(d_{PtPt}, t_0) \leq \frac{1}{B} P^{eq}_{GS}(d_{PtPt}) \end{cases} \qquad (10.6)$$

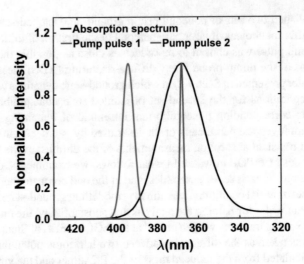

Fig. 10.2 Spectral intensity profiles of the ultrashort laser pulses used to generate initial conditions for two distinct sets of ground- and excited-state QM/MM BOMD nonequilibrium simulations. "Pump pulse 1" is the pump laser that we have used in the time-resolved XDS XFEL experiment on PtPOP in water [13, 14] (see Sect. 7.2). "Pump pulse 2" is the pump laser utilized by van der Veen et al. [12] in femtosecond optical measurements to investigate the vibrational relaxation of PtPOP in the first singlet excited state in aqueous solution. Also shown is the absorption spectrum of PtPOP measured in water [13]

with $B > 1$, and where the normalization factor for $P'_{ES}(d_{PtPt}, t_0)$, given by $\int P'_{ES}(d_{PtPt}, t_0) d(d_{PtPt})$, represents the simulated excitation fraction. The parameters A and B defining the form of $P'_{ES}(d_{PtPt}, t_0)$ were chosen such to deliver a simulated excitation fraction close to the estimated experimental α and most closely match the initial position of the ground-state hole as obtained from a fit of the experimental data (the results of the fit will be shown in Chap. 12). Then, 50 S_1 ΔSCF-QM/MM trajectories were started from ground-state QM/MM BOMD configurations reflecting the distribution $P'_{ES}(d_{PtPt}, t_0)$. The S_1 trajectories were collected with a time step of 2 fs, and keeping the thermostat applied only to the solvent molecules. In total, the trajectories amounted to around 200 ps of ΔSCF-QM/MM BOMD data. The adequacy of the approximation of using harmonic potentials in Eq. (10.5) was ascertained by comparing with $P'_{ES}(d_{PtPt}, t_0)$ calculated using free energy surfaces obtained as Morse-potential fits to the potential of mean force (PMF) from the QM/MM BOMD simulations. The PMF were calculated from the S_0 and S_1 QM/MM BOMD sets of data according to:

$$w^s(d_{PtPt}) = -k_b T \ln(g^s_{PtPt}(d_{PtPt})) \tag{10.7}$$

where $T = 300$ K and $g^s_{PtPt}(d_{PtPt})$ is the pairwise Pt–Pt RDF obtained from the ground- or excited-state simulations. The complete procedure that we followed in order to obtain $P'_{ES}(d_{PtPt}, t_0)$ using the PMF was:

1. Compute $g^{GS,eq}_{PtPt}(d_{PtPt})$ from the ~460 ps of ground-state QM/MM BOMD data.

2. Calculate $w^{GS,eq}(d_{PtPt})$ from $g_{PtPt}^{GS,eq}(d_{PtPt})$ using Eq. 10.7.
3. Compute $g_{PtPt}^{ES}(d_{PtPt})$ from the $\sim$200 ps of ΔSCF-QM/MM BOMD data.
4. Compute $g_{PtPt}^{ES,eq}(d_{PtPt})$ from the $\sim$200 ps of ΔSCF-QM/MM BOMD data minus the first (nonequilibrated) 2.5 ps of each trajectory.
5. Check that $g_{PtPt}^{ES}(d_{PtPt})$ is characterized by an average Pt–Pt distance and width that are the same as those of $g_{PtPt}^{ES,eq}(d_{PtPt})$, the two RDFs differing only by the level of statistical noise. This benchmark justifies the use of $g_{PtPt}^{ES}(d_{PtPt})$ instead of $g_{PtPt}^{ES,eq}(d_{PtPt})$ to obtain the PMF of the excited state.
6. Calculate $w^{ES}(d_{PtPt})$ from $g_{PtPt}^{ES}(d_{PtPt})$ using Eq. 10.7.
7. Shift $w^{ES}(d_{PtPt})$ such that the energy difference with respect to $w^{GS,eq}(d_{PtPt})$ at the minimum of $w^{GS,eq}(d_{PtPt})$ is equal to 3.35 eV (the position of the maximum of the $S_0 \rightarrow S_1$ band of the experimental absorption spectrum).
8. Compute $P'_{ES}(d_{PtPt}, t_0)$ from Eqs. (10.4), (10.5) and (10.6) using the difference between $w^{ES}(d_{PtPt})$ and $w^{GS,eq}(d_{PtPt})$.

A negligible difference was found between the $P'_{ES}(d_{PtPt}, t_0)$ distributions obtained under the two approximations (use of harmonic potentials versus use of PMF from the QM/MM BOMD simulations). Indeed, the two PMF are quite harmonic near the minimum of the S_1 surface (see Fig. 10.3), and, moreover, the Pt–Pt distances of the

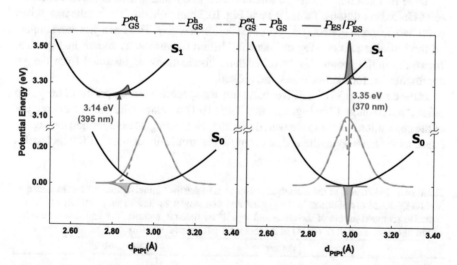

Fig. 10.3 Initial conditions for the $S_0 \rightarrow S_1$ photoexcitation of PtPOP in water simulated using the classical SF approximation with the pump pulse used in our ultrafast XDS measurements [13, 14] (Left), and with the pump pulse employed by van der Veen et al. [12] in transient absorption measurements in aqueous solution. The black curves are Morse-potential fits to the PMF calculated using the pairwise Pt–Pt RDFs obtained from the equilibrium QM/MM BOMD data for the ground state and from the first set of ΔSCF-QM/MM trajectories for the excited state, as explained in the text. A definition of the distributions appearing in the figure is provided in the text. The distributions were smoothed with a cubic smoothing spline

S_0 and S_1 gas-phase optimized geometries are found to be very close to the positions of the minima of the PMF (compare Fig. 10.3 with the values reported in Table 9.2).

The second set of initial conditions was chosen with a view to modelling the ultrafast vibrational relaxation in solution of a non-stationary ensemble of PtPOP molecules in the S_1 state. To achieve this we employed the parameters of the excitation pulse used in the transient absorption setup by van der Veen et al. [12] to probe the ultrafast excited-state dynamics of the complex in water. ω_1 corresponds approximately to the position of the maximum of the experimental absorption spectrum, thus it gives a $\hbar\omega_1$ of 3.35 eV ($\sim$370 nm); while τ is 60 fs. The spectral intensity profile of the pulse is reported in Fig. 10.2. For this set of simulations we could employ the Morse potentials obtained from a fit to the PMF of equilibrium QM/MM BOMD ground- and excited-state ensembles, as described before. Again, the two potentials where shifted such that the energy difference at the minimum of the ground-state potential was equal to 3.35 eV (which, in this case, corresponds to the center frequency of the excitation pulse). Since the experimental excitation fraction is not known, we chose $A = 1$ in Eq. (10.5). Thereafter, $P_{ES}(d_{PtPt}, t_0)$ was obtained directly from Eqs. (10.4) and (10.5), and it was used to start 49 S_1 ΔSCF-QM/MM trajectories. The time step for the propagation was the same as that used in the first set of simulations.

Table 10.1 summarizes the parameters of the pump pulses used in the two different sets of initial conditions. Table 10.1 and Fig. 10.2 highlight the fact that the two pulses cover two different (narrow) ranges of excitation energies. This has a profound impact on the initial conditions for the nonequilibrium dynamics, as shown in Fig. 10.3, where the initial ground- and excited-state distributions, as obtained from the SF approximation for the two cases, are plotted.

In the case of the XDS XFEL experiment, the spectral intensity profile of the pump pulse ("pump pulse 1" in Fig. 10.2 and Table 10.1) overlaps with the low energy tail of the absorption spectrum measured in water. Neglecting nuclear motion during the pulse, as we have done, the ultrashort pulse electronically excites PtPOP molecules

Table 10.1 Parameters of the excitation pulses used to generate initial conditions for the nonequilibrium QM/MM distributions in the ground and first singlet excited states of PtPOP. $\epsilon(t)$ is the Gaussian temporal profile of the pulse and $|\epsilon(\omega)|^2$ its spectral intensity. $\Delta\tau$ and $\Delta\omega$ are the full widths at half maximum (FWHM) of the temporal and spectral intensity profiles, respectively

	Pump pulse 1[a]	Pump pulse 2[b]
$\epsilon(t)$		
τ (fs)	20	60
$\Delta\tau$ (fs)	47	140
$\hbar\omega_1$ (eV)	3.14 (395 nm)	3.35 (370 nm)
$|\epsilon(\omega)|^2$		
$\hbar\Delta\omega$ (eV)	0.052 (6.6 nm)	0.018 (2.0 nm)

[a] Pulse used in our time-resolved XDS experiment [13, 14]
[b] Pulse used in the transient absorption measurements by van der Veen et al. [12]

with a Pt–Pt distance at which the $S_0 \rightarrow S_1$ energy gap is resonant with the excitation energy. Given the shape and relative position of the S_0 and S_1 potentials of PtPOP, this means that the laser is able to excite only ground-state molecules with a short Pt–Pt distance, close to the position of the minimum of the S_1 potential, as can be seen from Fig. 10.3 (Left). The implication is that the initial distribution prepared in the excited state is vibrationally "cold", since it comprises PtPOP molecules with d_{PtPt} close to the excited-state Pt–Pt equilibrium distance. Therefore, the classical ensemble of S_1 trajectories is expected to exhibit little vibrational dynamics. In the ground state, on the other hand, the excitation window $F(d_{PtPt})$ of Eq. (10.4) is sufficiently narrow to burn a localized hole at short distances in the equilibrium ground-state distribution of Pt–Pt distances. Classically, we expect that the hole will show large amplitude motion following the excitation event, as an effect of the remaining (non-stationary) ground-state molecules equilibrating in the S_0 potential.

The pump laser used in the optical pump-probe experiments performed by van der Veen et al. [12] ("pump pulse 2" in Fig. 10.2 and Table 10.1) covers a range of excitation energies around the maximum of the absorption spectrum of PtPOP. Therefore, it is able to preferentially excite ground-state molecules close to the bottom of the S_0 potential. The excited-state distribution created initially by the pulse is out-of-equilibrium with respect to the minimum of the S_1 surface, while the ground-state hole is centered around the equilibrium Pt–Pt distance. Thus, in this case, we expect the dynamics of the total ensemble (ground- plus excited-state molecules) to be dominated by coherent motion in the S_1 state (as we will see, the ground-state hole still exhibits a periodic spreading and refocusing, but no coherent vibrations).

The results of the two sets of simulations will be presented, separately, in great detail in Chaps. 12 and 13.

10.2.1 Considerations on the SF Approximation

In the following, we will briefly discuss what assumptions are made on the laser excitation process within the spatial filtering (SF) approximation (Eqs. 10.4 and 10.5). We will focus, in particular, on the use of fixed potential energy surfaces when applying the SF approximation in the context of QM/MM BOMD simulations, presenting evidence that an alternative method for simulating the transition based on a match between the instantaneous energy gap and the photon energy does not necessarily provide more reliable initial conditions for the nonequilibrium dynamics.

The SF approximation has been frequently employed to describe the pump-pulse transition in simulations of ultrafast pump-probe experiments using classical trajectories [7–11]. It is derived from first-order time-dependent perturbation theory under two main approximations:

- The transition dipole moment is assumed independent of the nuclear coordinates of the system in the Franck Condon region (Condon approximation).

- All kinetic energy operators in the expression of the promoted excited-state wave function are neglected [7, 9, 10], which means neglecting nuclear motion during laser excitation. We will consider the effects of nuclear motion during the pulse on the final outcome of a pump-probe experiment in Chap. 12, when we will discuss the optimal pump-pulse parameters for highlighting ground-state dynamics.

In addition, fixed potential energy surfaces for the ground and excited states are used to construct the potential energy difference entering the expression of the excitation window (see Eq. 10.5), which implies assuming a one-to-one correspondence between nuclear coordinates and the potential energy difference. The use of fixed potential energy surfaces appears to a good extent justified for diatomic molecules in gas phase or solid matrix, which have been the most common subject of classical molecular dynamics investigations that used the SF approximation [7, 9, 10]. On the other hand, in the case of polyatomic molecules in solution, it is clear that fluctuations in the surrounding environment and motion along other coordinates modify instantaneously the potentials along a particular coordinate [8].

We have investigated this effect by analysing the correlation between the Pt–Pt distance in PtPOP and the $S_0 \rightarrow S_1$ vertical transition energy in our QM/MM BOMD simulations. To do so, we have computed the $S_0 \rightarrow S_1$ energy gap for around 110000 QM/MM snapshots from the $\sim$460 ps of 300 K equilibrated ground-state QM/MM BOMD data by performing single-point ΔSCF-QM/MM calculations. The underlying distribution of instantaneous energy gaps is shown in the right panel of Fig. 10.4. The solution average $S_0 \rightarrow S_1$ transition energy is equal to 3.24 eV. This value is only slightly smaller than the computed $S_0 \rightarrow S_1$ vertical excitation energy in vacuum (3.51 eV, see Table 9.1) and within $\sim$4% of the position of the maximum of the room-temperature absorption spectrum of PtPOP in water ($\sim$3.35 eV [15, 18]). Figure 10.4 shows the joint probability distribution of Pt–Pt distances and $S_0 \rightarrow S_1$ energy gaps obtained from the 110000 single-point ΔSCF-QM/MM calculations. As expected, given the relative position of the PMF of S_0 and S_1 (see Fig. 10.3), there is a clear tendency towards higher excitation energies for larger Pt–Pt distances. However, the correlation coefficient computed from the covariance of the two variables, Pt–Pt distance and energy gap, is equal to 0.62, meaning that there is no sharp one-to-one correlation between the two parameters. This is further underpinned by the fact that the mean excitation energy for a given Pt–Pt distance (black line in Fig. 10.4) deviates from the mean Pt–Pt distance for a given energy gap value (red line).

There is an alternative strategy to the SF approximation for describing the pump-pulse transition in order to generate initial conditions for nonequilibrium classical distributions. The method consists in selecting ground-state configurations for which the instantaneous ground to excited state energy gaps are resonant with the photon energies of the excitation laser (i.e. according to the classical limit of the Franck-Condon principle [19]). This alternative approach has been used in the past to create the initial conditions for excited-state MD simulations aimed at modelling ultrafast pump-probe experiments on diatomics in solid rare gases [20, 21] or bond dynamics in liquids [8, 22, 23]. In those studies, first a set of ground-state equilibrated trajectories was established, and then configurations to excite from this set were chosen

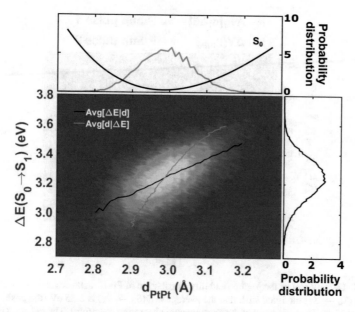

Fig. 10.4 Density plot of the joint probability distribution of Pt–Pt distances and $S_0 \rightarrow S_1$ energy gaps showing the correlation between these two parameters in the QM/MM BOMD simulations. The vertical excitation energies were computed by around 110000 single-point ΔSCF-QM/MM calculations on QM/MM BOMD ground-state configurations. The red and black lines superimposed to the bivariate distribution are the mean energy for a given Pt–Pt distance, and the mean Pt–Pt distance for a given energy gap, respectively. The panels at the right and upper sides of the density plot show the projection of the distribution along the energy gap and along the Pt–Pt distance (red curve), respectively. The upper panel includes, additionally, the PMF of the ground state (black curve)

according to a match with the energies within the bandwidth of the excitation pulse. In most of the cases, the classical propagation and the determination of the resonance condition were based on fixed potential energy surfaces. Therefore, the approach employed in most of those studies was very similar to the SF approximation. During our on-the-fly QM/MM BOMD simulations of PtPOP in water, the potential energy along the Pt–Pt distance adjusts instantaneously to the environment. Due to the lack of one-to-one correlation between the Pt–Pt distance and the $S_0 \rightarrow S_1$ transition energy, as seen before, the result of a selection based on the match between the laser energy and the $S_0 \rightarrow S_1$ energy gap is expected to be different from the outcome of the SF approximation. To test this, we have determined the S_1 distribution of Pt–Pt distances ($P_{ES}(d_{PtPt}, t_0)$) created by the ultrashort lasers used in the previous section within the SF approximation (see Table 10.1) by selecting PtPOP ground state molecules according to their values of instantaneous energy gap. The fixed potentials employed within the SF approximation have been shifted relative to each other such that the energy difference at the minimum of the ground-state potential was equal to the position of the maximum of the $S_0 \rightarrow S_1$ band of the experimental absorption

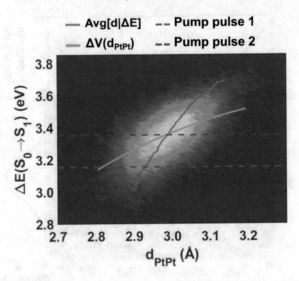

Fig. 10.5 Density plot of the joint probability distribution of Pt–Pt distances and $S_0 \rightarrow S_1$ energy gaps $\Delta E(S_0 \rightarrow S_1)$ modified such that the average $\Delta E(S_0 \rightarrow S_1)$ is 3.35 eV (the position of the maximum of the $S_0 \rightarrow S_1$ band of the experimental absorption spectrum). The red and green lines superimposed to the bivariate distribution are the mean Pt–Pt distance for a given energy gap and the difference between the S_0 and S_1 PMF of PtPOP along the Pt–Pt distance ($\Delta V(d_{PtPt})$), respectively. The dashed vertical lines represent the center energies of the pump pulses considered in the present investigation (pump-pulse parameters are reported in Table 10.1)

spectrum (3.35 eV). Thus, in order to enable a comparison between the two different strategies to draw initial conditions, the distribution of computed $S_0 \rightarrow S_1$ energy gaps $\Delta E(S_0 \rightarrow S_1)$ was modified such to give an average $\Delta E(S_0 \rightarrow S_1)$ of 3.35 eV. The bivariate distribution of Pt–Pt distances and modified energy gaps is shown in Fig. 10.5. Note how the center of the modified bivariate distribution is at 3.35 eV. Then, we computed $P_{ES}(d_{PtPt}, t_0)$ for the two laser pulses of Table 10.1 by first multiplying the spectral intensity profile of the laser with the bivariate distribution of Pt–Pt distances and modified energy gaps, and then projecting the resulting bivariate distribution along the Pt–Pt distance. For excitation with "pump pulse 1" (see Table 10.1), we scaled the intensity of the pulse to achieve the experimental excitation fraction, as also done when we have used the SF approximation with this pulse (see previous section); while, there was no need to rescale the resulting $P_{ES}(d_{PtPt}, t_0)$ according to Eq. (10.6), because excitation did not depopulate entirely the ground-state equilibrium distribution at particular Pt–Pt distances, in this case. The results obtained using the two pulses are shown in Fig. 10.6, together with the excited-state initial distributions computed with the SF approximation, and already presented in Fig. 10.3.

A first, most noticeable difference between the S_1 Pt–Pt distance distributions $P_{ES}(d_{PtPt}, t_0)$ created using the two different strategies (SF approximation and selection of instantaneous energy gaps according to the resonance condition) that emerges

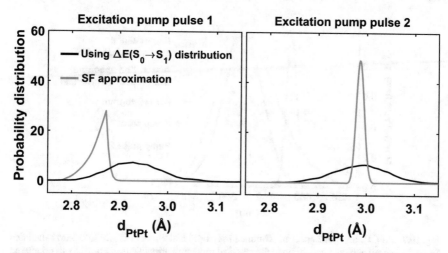

Fig. 10.6 Normalized distributions of Pt–Pt distances created in the S_1 state ($P_{ES}(d_{PtPt}, t_0)$) using two different approaches to describe the excitation of PtPOP by two ultrashort laser pulses. The black curves were obtained by filtering the bivariate ground-state distribution of Pt–Pt distances and $S_0 \rightarrow S_1$ energy gaps with the spectral intensity profiles of the lasers. The red curves were obtained within the SF approximation (see Eqs. (10.4) and (10.5), and Fig. 10.3), using the difference between the S_0 and S_1 PMF of PtPOP along the Pt–Pt distance ($\Delta V(d_{PtPt})$). See Table 10.1 for the parameters of the pump pulses

from Fig. 10.6, is that the distributions created according to the second strategy are much broader than the one obtained with the SF approximation. The reason for this result is understood by observing from Fig. 10.5 that the range of Pt–Pt distances at the center energies of the pump pulses (dashed horizontal lines) are very disperse due to the lack of perfect correlation between the Pt–Pt distance and the vertical transition energy. Hence, at the energies covered by the laser pulses, the PtPOP molecules that are eligible to be selected for excitation according to the resonance condition will display a large distribution of Pt–Pt distances. On the other hand, the SF approximation filters the ground-state equilibrium distribution of Pt–Pt distances according to a d_{PtPt}-dependent window function (see again Eqs. 10.4 and 10.5), and thus selects molecules in configuration space based on the fixed potential energy difference, regardless of the instantaneous energy gap. In this latter picture, the excited distributions will have a broad width in the space of $S_0 \rightarrow S_1$ transition energies. A second difference between the predictions of the two approaches is represented, in the case of "pump pulse 1", by a shift of the position of the distribution excited using the instantaneous energy gaps to shorter Pt–Pt distances, closer to the value of the average Pt–Pt distance of the ground state (2.99 Å), with respect to $P_{ES}(d_{PtPt}, t_0)$ predicted by the SF approximation. To understand the origin of the discrepancy, we have plotted, additionally, in Fig. 10.5 the mean Pt–Pt distance for a given value of energy gap (red line) and the difference between the S_1 and S_0 PMF used within the SF approximation ($\Delta V(d_{PtPt})$, green line). The intersection between the red line and

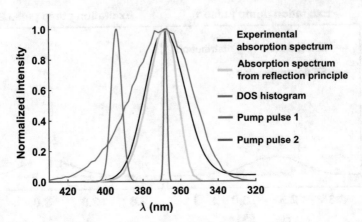

Fig. 10.7 PtPOP absorption spectrum simulated as a DOS histogram of ΔSCF-QM/MM calculated $S_0 \rightarrow S_1$ energy gaps and according to the classical reflection principle (see Eq. 10.8), as compared to the experimental spectrum of PtPOP in water [13]. Also shown are the Gaussian spectral intensity profiles of the laser pulses used in the present work to set up initial conditions for nonequilibrium dynamics of PtPOP in water

the horizontal dashed line representing the center energy of "pump pulse 1" (3.14 eV) gives the center of the distribution of Pt–Pt distances that can be excited by the laser, in the picture based on the selection of instantaneous energy gaps. While in the case of the SF approximation, the molecules that are promoted to the excited state have a Pt–Pt distance close to the intersection between the horizontal dashed line and the green line representing $\Delta V(d_{PtPt})$. Thus, it is apparent that, due to the disruption of the correlation between $S_0 \rightarrow S_1$ energy gaps and Pt–Pt distances, the average Pt–Pt distance at the energy of the pulse is closer to the average ground-state Pt–Pt distance than the Pt–Pt distance where $\Delta V(d_{PtPt})$ is equal to the pulse energy, explaining the shift of the position of $P_{ES}(d_{PtPt}, t_0)$ for "pump pulse 1". For "pump pulse 2", instead, the resonance condition is satisfied at the center of the ground-state distribution, where $\Delta V(d_{PtPt})$ and the red line coincide. As a consequence, the excited Pt–Pt distance distributions are centered both at the average Pt–Pt distance of the ground state, for the two different method of selecting initial conditions.

A question that arises now is which one of the two strategies for setting up the initial conditions for the nonequilibrium dynamics provides a better approximation to the pump-pulse transition. To shed light on this question, we compare in Fig. 10.7 the experimental absorption spectrum of PtPOP in water [13] to the spectra simulated as a density-of-states (DOS) histogram from the (modified) distribution of $S_0 \rightarrow S_1$ energy gaps and using the classical reflection principle [24]. The classical reflection principle states that the energy dependence of the probability of absorption of a photon reflects the equilibrium nuclear distribution in the ground state in a direct way, and is valid within the same set of assumptions as those used to derive the

SF approximation. The absorption cross section of PtPOP was computed, according to the classical reflection principle, from the 300 K equilibrium ground-state distribution $P_{GS}^{eq}(d_{PtPt})$, using the following relation:

$$\sigma(E) \propto \frac{1}{|\Delta V'(d_{PtPt})|} P_{GS}^{eq}(d_{PtPt}) \tag{10.8}$$

where $\Delta V'(d_{PtPt})$ is the derivative of the difference potential between ground and excited states, which was obtained from the S_0 and S_1 PMF of PtPOP along the Pt–Pt coordinate. The spectrum calculated in this way gives the probability of absorption at the different Pt–Pt distances within the SF approximation.

There is a good agreement between the spectrum simulated using the reflection principle and the experimental spectrum, especially on the red side, while at higher energies, for which the underlying potentials are more anharmonic, the experimental spectrum is broader. On the other hand, the DOS histogram clearly overestimates the width of the spectrum over the entire range of energies. The distribution of instantaneous $S_0 \rightarrow S_1$ energy gaps, from which the DOS histogram is constructed, was obtained from single point ΔSCF-QM/MM calculations on ground-state QM/MM BOMD configurations. In these calculations, the geometry of the solute and all solvent molecules are kept fixed (frozen-field approximation). It can be argued, however, that the Pt–Pt oscillators should only experience an effective field during the rapid fluctuations induced by the environment [25]. The failure of the frozen-field approximation to correctly describe such phenomena of motional narrowing that are active in solution [8, 25] provides an explanation of the discrepancy between the experimental absorption spectrum and the spectrum simulated as a DOS histogram.

Figure 10.7 shows also the spectral intensity profiles of the two laser pulses whose parameters were utilized in the present study to set up initial conditions for the nonequilibrium dynamics of PtPOP in water. It is seen that, at the range of energies covered by the pulses, the spectrum predicted by the classical reflection principle is in very good agreement with the experimental absorption spectrum. It can be argued, on the basis of this comparison, that the ground-state equilibrium distribution $P_{GS}^{eq}(d_{PtPt})$ directly reflects via the potential difference $\Delta V(d_{PtPt})$ the probability of photon absorption over the range of energies covered by the pump pulses. Therefore, using $P_{GS}^{eq}(d_{PtPt})$ and $\Delta V(d_{PtPt})$ within the SF approach, as done in the present work, represents a reasonable approximation of the excitation process. On the other hand, due to effects of motional narrowing, a direct correlation between the distribution of instantaneous energy gaps calculated in the frozen-field approximation and the absorption spectrum is difficult to establish. Thus, a selection of initial QM/MM BOMD configurations based solely on a match of the energy gap between ground and excited state with the resonant energies of the excitation laser does not necessarily imply an improvement of the accuracy of the initial conditions.

References

1. Jorgensen WL (1981) Quantum and statistical mechanical studies of liquids. 10. transferable intermolecular potential functions for water, alcohols, and ethers. Application to liquid water. J Am Chem Soc 103:335–340
2. Jensen KP, Jorgensen WL (2006) Halide, ammonium, and alkali metal ion parameters for modeling aqueous solutions. J Chem Theory Comput 2(6):1499–1509
3. Larsen AH, Vanin M, Mortensen JJ, Thygesen KS, Jacobsen KW (2009) Localized atomic basis set in the projector augmented wave method. Phys Rev B 80:195112
4. Rappe AK, Casewit CJ, Colwell KS, Goddard WA III, Skiff WM (1992) Uff, a full periodic table force field for molecular mechanics and molecular dynamics simulations. J Am Chem Soc 114:10024–10035
5. Andersen HC (1983) Rattle: a "velocity" version of the shake algorithm for molecular dynamics calculations. J Comput Phys 52:24
6. Frenkel D, Smit B (2002) Understanding molecular simulation. Academic Press
7. Petersen J, Henriksen NE, Møller KB (2012) Validity of the Bersohn-Zewail model beyond justification. Chem Phys Lett 539–540:234–238
8. Møller KB, Rossend R, Hynes JT (2004) Hydrogen bond dynamics in water and ultrafast infrared spectroscopy: a theoretical study. J Phys Chem A 108:1275–1289
9. Ermoshin VA, Engel V (2001) Femtosecond pump-probe fluorescence signals from classical trajectories: comparison with wave-packet calculations. Eur Phys J D 15:413–422
10. Li Z, Fang J-Y, Martens CC (1996) Simulation of ultrafast dynamics and pump-probe spectroscopy using classical trajectories. J Chem Phys 104(18):6919
11. Jonas DM, Bradforth SE, Passino SA, Fleming GR (1995) Femtosecond wavepacket spectroscopy: influence of temperature, wavelength, and pulse duration. J Phys Chem 99(9):2594–2608
12. van der Veen RM, Cannizzo A, van Mourik F, Vlček Jr A, Chergui M (2011) Vibrational relaxation and intersystem crossing of binuclear metal complexes in solution. J Am Chem Soc 113:305
13. Haldrup K, Levi G, Biasin E, Vester P, Laursen MG, Beyer F, Kjær KS, Brandt van Driel T, Harlang T, Dohn AO, Hartsock RJ, Nelson S, Glownia JM, Lemke HT, Christensen M, Gaffney KJ, Henriksen NE, Møller KB, Nielsen MM (2019) Ultrafast x-ray scattering measurements of coherent structural dynamics on the ground-state potential energy surface of a diplatinum molecule. Phys Rev Lett 122:063001
14. Biasin E, van Driel TB, Levi G, Laursen MG, Dohn AO, Moltke A, Vester P, Hansen FBK, Kjaer KS, Hartsock R, Christensen M, Gaffney KJ, Henriksen NE, Møller KB, Haldrup K, Nielsen MM (2018) Anisotropy enhanced X-ray scattering from solvated transition metal complexes. J Synchrotron Radiat 25(2):306–315
15. Stiegman AE, Rice SF, Gray HB, Miskowski VM (1987) Electronic spectroscopy of d^8-d^8 diplatinum complexes. $^1a_{2u}(d\sigma^* \to p\sigma)$, $^3e_u(d_{xz},d_{yz} \to p\sigma)$, and $^{3,1}b_{2u}(d\sigma^* \to d_{x^2-y^2})$ excited states of $pt_2(p_2o_5h_2)_4^{4-}$. Inorg Chem 26:1112
16. Rice SF, Gray HB (1983) Electronic absorption and emission spectra of binuclear platinum(II) complexes. Characterization of the lowest singlet and triplet excited states of $Pt_2(P_2O_5H_2)_4^{4-}$. J Am Chem Soc 105:4571–4575
17. Van Kleef EH, Powis I (1999) Anisotropy in the preparation of symmetric top excited states. I. One-photon electric dipole excitation. Mol Phys 96(5):757–774
18. Fordyce WA, Brummer JG, Crosby GA (1981) Electronic spectroscopy of a diplatinum(II) octaphosphite complex. J Am Chem Soc 103(6):7061–7064
19. Franck J, Dymond EG (1926) Elementary processes of photochemical reactions. Trans Faraday Soc 21:536–542
20. Batista VS, Coker DF (1997) Nonadiabatic molecular dynamics simulation of ultrafast pump-probe experiments on I_2 in solid rare gases. J Chem Phys 106:6923

21. Zadoyan R, Li Z, Martens CC, Apkarian VA (1994) The breaking and remaking of a bond: Caging of I2 in solid Kr. J Chem Phys 101(8):6648–6657
22. Dohn AO, Jónsson EÖ, Kjær KS, van Driel TB, Nielsen MM, Jacobsen KW, Henriksen NE, Møller KB (2014) Direct dynamics studies of a binuclear metal complex in solution: the interplay between vibrational relaxation, coherence, and solvent effects. J Phys Chem Lett 5:2414–2418
23. Winter N, Benjamin I (2004) Photodissociation of ICN at the liquid/vapor interface of water. J Chem Phys 121(5):2253–2263
24. Schinke R (1993) Photodissociation dynamics. Cambridge University Press
25. Ojamäe L, Tegenfeldt J, Lindgren J, Hermansson K (1992) Simulation of band widths in liquid water spectra. The breakdown of the frozen-field approximation. Chem Phys Lett 195(1):97–103

Chapter 11
Equilibrium Solution Structure

11.1 PtPOP Equilibrium Structure

Table 11.1 reports bond lengths and angles of PtPOP obtained as averages over thermally equilibrated S_0 and S_1 QM/MM BOMD data in water. The ground-state equilibrium QM/MM BOMD simulations were described in Sect. 10.1. Equilibrium data for S_1 were extracted from the two sets of excited-state trajectories (see Sect. 10.2) after removing the first (nonequilibrated) 2.5 ps from each of them, which gave a total of around 80000 BOMD snapshots, covering 160 ps. That the molecule vibrationally equilibrates after the first 2.5 ps during the S_1 QM/MM BOMD simulations will be shown in Chap. 13.

The only structural parameter of the S_1 state of PtPOP in aqueous solution that has ever been determined experimentally is the equilibrium Pt–Pt distance. This bond length has been obtained in the course of the present PhD project by E. Biasin [5] from a fit to the time-dependent XDS difference scattering signal presented in Sect. 7.2 (the data are shown in Fig. 7.3). For this analysis, the signal was fitted at a pump-probe time delay of 4.5 ps, when both the ground- and excited-states of PtPOP have reached vibrational equilibrium. In Sect. 11.3 of the present chapter, we will have a closer look at how the simulations that we have performed helped guiding the analysis of the data. The experimentally determined Pt–Pt contraction for S_1 was found to be equal, within the accuracy of the experiments, to that obtained by time-resolved X-ray scattering measurements in water by Christensen et al. [3] for the T_1 state. Based on this, and given the very close similarity between the S_1 and T_1 vacuum structures (see Table 9.2), we test the same holds in solvent, namely that the solution structures of the S_1 and T_1 states are virtually the same. Thus, in Table 11.1 we carry out a comparison between calculated thermal averages for S_1 and corresponding experimental solution data available for T_1. These include the Pt–Pt and P$\cdots$P$'$ distances measured in the aforementioned X-ray scattering experiment performed by Christensen et al. [3], and the Pt-P bond lengths derived by van der Veen et al. [4, 6] from a fit to time-resolved X-ray absorption spectra in ethanol.

© Springer Nature Switzerland AG 2019

G. Levi, *Photoinduced Molecular Dynamics in Solution*, Springer Theses,
https://doi.org/10.1007/978-3-030-28611-8_11

Table 11.1 Structural parameters of PtPOP in water obtained as averages over equilibrium QM/MM BOMD data for the S_0 and S_1 states, and comparison with available solution experimental values[a]. The MD average was carried out over a total simulation time of $\sim$460 ps for S_0 and of $\sim$160 ps for S_1

	S_0		$\Delta (S_1 - S_0)$		$\Delta (T_1 - S_0)$
	Calc [2]	Exp [3, 4]	Calc	Exp [5]	Exp [3, 6]
Bond (Å)					
Pt–Pt	2.99	2.98[b]	−0.20	−0.24(4)[b]	−0.24(6)[b]
Pt–P	2.33	2.32(4)[c]	0.01	–	0.010(6)[c]
P$\cdots$P′	3.09	2.92[b]	−0.01	–	0.00(8)[b]
Angles (deg)					
(Pt–Pt–P)$_\alpha$	91.2	–	5.0	–	–
(Pt–Pt–P)$_\beta$	91.2	–	−0.3	–	–

[a] Simulation results for S_1 are compared to experimental values obtained for T_1 when experimental data for S_1 are not available
[b] Obtained in water by X-ray scattering experiments [3, 5]
[c] Obtained in ethanol by X-ray absorption measurements [4, 6]

The calculated thermally averaged bond distances and the experimental values agree within the uncertainties of both experiments and simulations. From a comparison with the corresponding structural parameters of the gas-phase optimized geometries, we notice that the average Pt–Pt distance in solution is only 0.01 Å shorter, while the solvent affects much more significantly structural parameters involving ligand atoms. This is particularly evident for the Pt-P bonds, which in the ground state are found to be $\sim$0.06 Å shorter than in the isolated optimized structure and in S_1 experience a $\sim$70% smaller elongation. In addition, despite the fact that the shortening of the Pt–Pt distance due to excitation is found to be the same in vacuum and solution, the P atoms follow the Pt atoms in the contraction along the Pt–Pt axis to only 0.01 Å, $\sim$80% less than in gas phase. As a side note, we point out that differences induced by the presence of the solvent on these structural parameters of PtPOP are larger than the changes brought by the use of a hybrid DFT functional like B3LYP, as can be seen by comparing the values reported for the Pt-P and P$\cdots$P′ distances in Tables 9.2 and 11.1. This indicates that there would be no significant advantage in employing the computationally more expensive B3LYP functional instead of BLYP in the QM/MM BOMD simulations.

An analysis of the average values of the $\angle$Pt–Pt–P angles reveals that also in solution, PtP$_4$ units are distorted towards a quasi-trigonal bipyramidal local geometry with respect to the ground state (though the angle difference Δ found in solvent is $\sim$5°, around 2° smaller than for the optimized S_1 vacuum geometry). This is an important result, because it hints at the fact that a direct $S_1 \rightarrow T_1$ ISC mechanism might be active in solution, which could explain the $\sim$3000-times faster ISC rates exhibited by PtPOP with respect to its perfluoroborated analogue [7, 8], where pseudorotation of the bulkier and more rigid ligands is less likely. Indeed, the role of structural

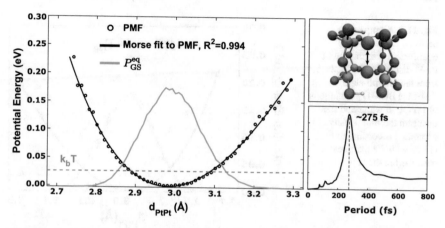

Fig. 11.1 (Left) Potential of mean force (PMF) calculated from the Pt–Pt radial distribution function (RDF) according to Eq. 10.7, together with the relative thermal probability distribution ($P_{GS}^{eq}(d_{PtPt})$) from the QM/MM BOMD equilibrium simulations of PtPOP in water. The red line defines the average thermal energy available to the system at 300 K. (Bottom, right) Fourier transform of the Pt–Pt oscillations in the simulations

distortions in lowering the D_{4h} symmetry of the Pt_2P_8 core of PtPOP, thus promoting direct SOC between S_1 and T_1, has been often hypothesized but so far never proven [7–10].

11.1.1 Thermal Equilibrium Properties of the Ground State

Here, we expand on the thermally averaged structural and dynamical properties of the Pt–Pt distance in the ground state. We show that the large amount of QM/MM BOMD data collected (∼460 ps) permits a statistically robust characterization of the distribution of Pt–Pt distances and frequency of the Pt–Pt oscillations in equilibrium with a thermal bath of solvent molecules at 300 K.

Figure 11.1 (Left) shows the 300 K distribution of Pt–Pt distances for the ground state of PtPOP ($P_{GS}^{eq}(d_{PtPt})$), together with the free-energy surface obtained as potential of mean force (PMF) (see Eq. 10.7). This equilibrium distribution has been used in Sect. 10.2 (see Fig. 10.3) to set up the initial conditions for the nonequilibrium BOMD simulations that will be presented in the next chapters. The thermally averaged Pt–Pt distance of PtPOP in aqueous solution from our simulations is 2.99 Å, while he experimental value obtained from X-ray scattering measurements in water is 2.98 Å [3]; therefore, the discrepancy between simulations and experiment is less than 1%.

The PMF shown in Fig. 11.1 (Left) has been fitted to a Morse potential, which has the form [11]:

$$V(d_{PtPt}) = D_e \left[1 - e^{-a\left(d_{PtPt} - d_{PtPt,0}\right)} \right]^2 \tag{11.1}$$

Fig. 11.2 Comparison
between the (classical)
equilibrium distribution of
Pt–Pt distances obtained
from the ground-state
QM/MM BOMD simulations
of PtPOP in water and the
quantum thermal density
computed according to Eq.
(11.2) with the parameters
described in the text

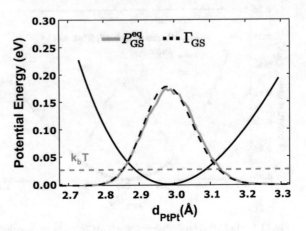

where D_e and $d_{PtPt,0}$ are the depth and Pt–Pt distance at the potential minimum, respectively, and $a = \sqrt{k_0/2D_e}$ with k_0 the force constant at the minimum of the potential well. Figure 11.1 (Left) reports the Morse potential resulting from the fit. We have constructed a quantum thermal density as an incoherent sum of vibrational eigenstates of this Morse potential, according to:

$$\Gamma_{GS} = \sum_{i=1}^{N_s} p_i \, |\chi_i\rangle \, \langle\chi_i| \tag{11.2}$$

where $|\chi_i\rangle$ are the first N_s eigenstates with eigenvalues ε_i, and the probabilities p_i are the Boltzmann factors for the canonical ensemble ($p_i = \dfrac{1}{\sum_i^{N_s} e^{-\beta\varepsilon_i}} e^{-\beta\varepsilon_i}$, with $\beta = (k_b T)^{-1}$).

In the computation of Γ_{GS} according to Eq. (11.2), N_s was chosen to include all states with a probability bigger than 0.1%, as determined using the first 250 eigenvalues, and the temperature T was 300 K, the same temperature set for the Langevin thermostat applied to the water molecules in the QM/MM BOMD simulations. The eigenstates and corresponding eigenvalues of the Morse potential were obtained using the Matlab program WavePacket [12]. Figure 11.2 shows a comparison between the equilibrium distribution of Pt–Pt distances obtained from the QM/MM BOMD simulations and the quantum density Γ_{GS} at 300 K. The comparison indicates that the classical probability distribution of Pt–Pt distances approximates very well the quantum density. From the comparison we can deduce that: (i) the QM/MM BOMD simulations correctly reproduce the fluctuations of the canonical ensemble, and (ii) at 300 K the predictions of classical statistical mechanics with respect to the Pt–Pt oscillators are close to the quantum-classical limit. To understand qualitatively the latter observation we can compute the energy spacing of a harmonic potential with a frequency given by the force constant of the Morse potential of Fig. 11.2 and the reduced mass μ_{PtPt} of Pt$_2$ ($\nu_{PtPt} = 1/2\pi\sqrt{k_0/\mu_{PtPt}}$). The frequency of the Morse potential is, in terms of wavenumbers, $\tilde{\nu}_{PtPt} = 124$ cm^{-1}. This gives $\beta h\nu_{PtPt} = 0.59$; and we can see that at 300 K we are already close to approach the high temperature limit (for a harmonic oscillator, classical and quantum statistical mechanics give the same prediction for $\beta h\nu_{PtPt} \ll 1$).

We now turn to examine the dynamical properties of the Pt–Pt oscillator in the ground state. From the Morse-potential fit to the PMF a low degree of anharmonicity of the Pt–Pt stretching vibration can be deduced. The degree of deviation from harmonicity was estimated by calculating the anharmonicity constant x_e according to the expression [11]:

$$x_e = \frac{h\nu}{4D_e} = \frac{\hbar}{4D_e}\sqrt{\frac{k_0}{\mu_{PtPt}}} \tag{11.3}$$

Using the parameters of the Morse-potential fit, a value of $x_e = 1.5 \cdot 10^{-3}$ is obtained, which is noticeably small if compared, for example, to that of a very harmonic diatomic system like I_2 ($2.8 \cdot 10^{-3}$ [13]). The position of the minimum of the Morse-fitted potential at 2.98 Å, a Pt–Pt distance only 0.01 Å shorter than the thermally averaged value, also points to a strong harmonicity. Finally, we have obtained the Pt–Pt vibrational frequency as the maximum of a Fourier transform (FT) of the oscillating Pt–Pt distance in the simulations. The FT is reported in Fig. 11.1 (Bottom, right). To obtain it, we have divided each trajectory into chunks of 4 ps, performed an FT for each of them, and then averaged the results. The oscillation periods obtained from the frequency of the Morse potential, reported before, and the position of the maximum of the FT are 270 and 275 fs, respectively. The small deviation between the two values is a further confirmation of the harmonicity of the Pt–Pt potential. We also note that the computed periods deviate by less than 4 % from the vibrational period of 281 fs obtained by van der Veen et al. [14] using femtosecond transient absorption measurements in aqueous solution. More importantly, for the purpose of the interpretation of the XDS XFEL data measured during the present project (see Sect. 7.2), the simulated Pt–Pt oscillation period is found to be very close to the period ($\sim$285 fs) obtained from the position of the maximum of an FT of the time-dependent XDS signal (see Fig. 7.3).

11.2 Solvation Shell Structure

Figure 11.3 shows the Pt-H$_{solvent}$ and Pt-O$_{solvent}$ radial distribution functions (RDFs) computed from the equilibrated QM/MM trajectories in S_1 and S_0 using a bin size of 0.01 Å for the radial sampling. The large amount of statistics allows to fully resolve the first four peaks of solvent coordination around the Pt atoms. The position of water molecules within each of the coordination peaks with respect to a single Pt atom is illustrated schematically in Fig. 11.3 (Right).

The first peak in the two RDFs (areas of the RDFs highlighted in blue in Fig. 11.3 (Left, bottom) are indicative of the presence of strong H-coordination of solvent

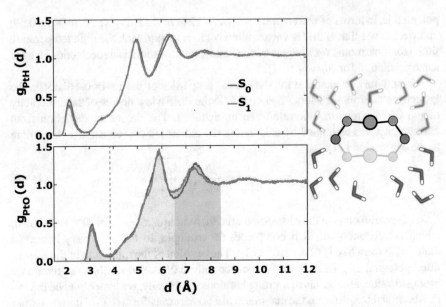

Fig. 11.3 (Left) Pairwise Pt-solvent RDFs sampled from the equilibrated part of the QM/MM trajectories in S_0 and S_1. The gray vertical line indicates the extent of the first coordination peak. (Right) Division of the solvation shell around PtPOP into regions from the point of view of a Pt atom. The colors of the different regions match the colors of the areas of the peaks of the Pt-$O_{solvent}$ RDFs

molecules at the Pt-ends of the complex. The second coordination peaks (yellow water molecules) span Pt-H $_{solvent}$ distances between ~4.5 and ~5.5 Å and Pt-$O_{solvent}$ distances between ~5.5 and ~6.5 Å. These peaks comprise water molecules that are found to lie mainly off-axis with respect to the Pt–Pt direction. Due to the presence of two Pt atoms, and given the symmetry of the complex, water molecules of the second peaks make up also the third peaks of the RDFs (5.5 Å $< d_{PtH} <$ 7.0 Å and 6.5 Å $< d_{PtO} <$ 8.3 Å). This is better illustrated by the water molecules highlighted in purple in the schematics of Fig. 11.3 (Right). That water molecules belonging to the second (and third) peaks do not take up the space along the Pt–Pt direction is supported by the fact that the distance between the second and third peaks (~1.5 Å) is less than the intramolecular Pt–Pt distance.

In the excited state, neither the position of the first nor the second peak changes, which means that the two shells must draw closer in conjunction with the Pt–Pt contraction. This is further supported by a shift of the third peaks to shorter distances as this results from the Pt atoms finding themselves closer to water molecules located on the opposite sides of the complex, in the excited state. The second most notable change in the RDFs is represented by a slightly reduced coordination in the first peaks. If the extent of the first coordination shell is taken up to the first minimum of the Pt-$O_{solvent}$ RDF, at 3.85 Å, we can quantify the coordination with the running coordination number at this distance. It follows that the Pt-O$_{solvent}$ coordination number in the first shell is around 0.77 for PtPOP in the S$_1$ state, only ~0.1 smaller than in the ground state.

11.2.1 Orientational Distribution in the First Coordination Peak

We have analysed the orientation of the water molecules in the first peak in the Pt-H$_{solvent}$ and Pt-O$_{solvent}$ RDFs in order to gain insight into the nature of the coordination at the open axial site of the complex.

Figure 11.4 shows the probability distribution function of two key angles in the solute-solvent geometry sampled within the first coordination peak. Both distributions indicate a preference for linear geometry, or axial coordination, where the O-H donor bond points along the Pt–Pt axis of the complex. This is further supported by the distance between the first peak of the Pt-H$_{solvent}$ RDF and that of the Pt-O$_{solvent}$ RDF, which is roughly 0.96 Å, corresponding to the TIP4P O-H bond length.

The angular distributions, indicative of the extent of axial coordination, are largely unaltered when PtPOP is in the S$_1$ excited state with respect to the ground state. Therefore, water molecules within the first peak of the RDFs seem to retain the preferential axial orientation after electronic excitation of the complex. Experimentally, emission spectra of PtPOP are found to be independent of the solvent [15]. The finding that electronic excitation does not lead to any major restructuring in the local organization of solvent molecules surrounding the complex is in agreement with this experimental observation and points to the fact that this might be the case also for other types of solvents.

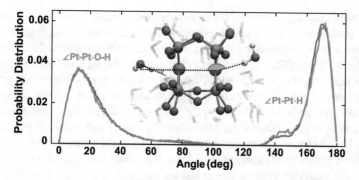

Fig. 11.4 Probability distributions of solute-solvent angles involving water molecules in the first coordination shell of the Pt atoms in PtPOP as defined by the extent of the first peak of the Pt-O$_{solvent}$ RDF in Fig. 11.3. The color code for the distributions is the same as in Fig. 11.3. The sampled angles are shown schematically using a QM/MM BOMD snapshot selected from one of the S$_1$ ΔSCF-QM/MM trajectories (for visualization purposes hydrogen atoms of PtPOP are omitted). These angles are related to the orientation of water molecules with respect to the Pt–Pt axis of the complex

This behaviour is in sharp contrast to the solvent shell response observed for photoexcitation of the d^8-d^8 complex [Ir$_2$(dimen)$_4$]$^{2+}$ (where dimen is diisocyano-para-menthane) by ultrafast X-ray scattering measurements in acetonitrile [16]. In that case, the effect of electronic excitation was found to be a loss of coordination of methyl groups with the open coordination site at the metal atoms, followed by reorientation of the solvent molecules to specifically coordinate Ir atoms with the more electronegative cyano endings. In both complexes a metal-metal bond is formed after photoexcitation, thus effectively shifting electron density from the outer side of the planar faces of the molecules to the inside (see Sect. 9.2). Although different solvents are involved in the two cases, the different response of coordinating solvent molecules can be rationalized in terms of different contributions of atomic orbitals localized on ligand atoms in the formation of the LUMO. As we have seen in Sect. 9.2, in PtPOP the LUMO has a largely predominant p_z character; as a consequence, in the excited state, a considerable portion of the electron density still localizes in outward position with respect to the planar PtP$_4$ faces. This, in turn, permits the Pt atoms of the complex to retain their ability to coordinate the more electropositive part of the solvent and it is probably connected to the previously mentioned excited-state reactivity towards H atom donors. For [Ir$_2$(dimen)$_4$]$^{2+}$, on the other hand, previous DFT calculations [7] have highlighted a substantial involvement of π_z^*(C≡N) orbitals in the formation of the LUMO, shifting more electron density from the outer sides of the molecule and making the excited state a stronger Lewis acid.

11.3   Guiding the Analysis of the XDS Data

11.3.1   Choosing a Structural Model for the Solute

The strategy outlined in Sect. 8.2 for analysing time-dependent difference X-ray scattering signals measured in solution reckons on a proper choice of single geometries to model the scattering due to changes in the distances between atoms of the solute (the $\Delta S^{solu}(q, t_p)$ term). The standard procedure used by the group of our experimental collaborators consists in utilizing DFT-optimized geometries. Geometry optimization is carried out in vacuum or by taking into account solvent effects using a continuum solvent model [16, 17]. To model structural changes after photoexcitation in the ground or excited state of the solute, usually sets of structures are generated by varying key structural parameters while optimizing at each step the geometry in the state where dynamics is expected [16]. Thus, the model approximates the scattering from time-dependent distributions of atomic positions within the ensemble of solvated solute molecules with the scattering of single geometries relaxed with respect to fixed PESs, those predicted by DFT in vacuum or in an implicit solvent model. Here, we test the choice of two types of vacuum structures for the analysis of the PtPOP XDS data.

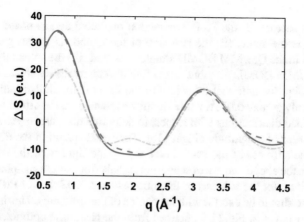

Fig. 11.5 Simulated isotropic difference scattering signal of PtPOP calculated using Eq. (8.5) from: the RDFs obtained from the S_0 and S_1 QM/MM equilibrium data (blue line), the optimized geometries of S_0 and S_1 (yellow line), and the optimized geometry of S_0 and an excited state described by the optimized geometry of S_0 with the Pt–Pt distance of the S_1 optimized geometry (red line). All geometry optimizations were performed in vacuum. The excitation fraction α in Eq. (8.5) was equal to 0.026. This value of α was determined at a later stage of the analysis

The first choice is akin to the one just illustrated: both ground and excited states of the complex are represented by the respective geometries optimized in vacuum using GPAW with the BLYP functional. The structural parameters of these geometries are reported in Table 9.2. DFT calculations with BLYP in vacuum predict a relatively large, $\sim$0.03 Å elongation of the Pt-P bond lengths, and a large, $\sim$0.06 Å contraction of the $P \cdots P'$ distances compared to the QM/MM BOMD simulations (see Table 11.1). On the other hand, the QM/MM BOMD simulations and the geometry optimizations in vacuum give the same value for the Pt–Pt contraction in the S_1 state. Therefore, a second choice of structures that we test is the following: the ground state is represented by the ground-state geometry optimized in vacuum in GPAW, while the excited state is obtained from the latter by varying exclusively the Pt–Pt distance, without relaxing the forces with respect to the S_1 PES.

Figure 11.5 shows the isotropic difference scattering signal calculated (i) using Eq. (8.5) with the solute-solute RDFs extracted from the S_0 and S_1 QM/MM BOMD equilibrium data (blue line); (ii) using the same equation but applied to the ground- and excited-state gas-phase optimized geometries of the complex, in which case Eq. (8.5) reduces to a sum of terms equivalent to the Debye formula Eq. (8.14), (yellow line); and (iii) using the gas-phase S_0 optimized geometry and an excited-state structure obtained from the first by setting the Pt–Pt distance to that of the optimized excited-state geometry, while leaving all other DOF unchanged (red line).

The comparison illustrates that the choice of a gas-phase excited-state structure with the same structural parameters as the ground state, apart from the Pt–Pt distance, approximates very well the scattering obtained from QM/MM BOMD thermal distributions. The result can be explained as the consequence of the convolution of at least three factors: (i) the Pt–Pt contraction predicted by gas-phase and solution calculations is the same, (ii) the structure of the ligands is unchanged in S_1 with respect to S_0 in the QM/MM BOMD simulations, and (iii) the spread that characterizes the QM/MM BOMD distributions is found to be relatively small. With respect to the last point, we note that the width of the Pt–Pt thermal distribution shown in Fig. 11.1 is only around 0.07 Å, while significant changes in the difference scattering from broadened distributions with respect to delta functions are expected for widths at least an order of magnitude bigger [18]. The small spread in the distributions is caused by the stiffness of the Pt–Pt bond and by the rigidity of the cage of ligand atoms. Given the above, the use of scattering signals from single gas-phase structures appears justified for the analysis of the time-resolved XDS data of PtPOP in water. However, one should be careful with the choice of the structures. This is exemplified by the yellow curve in Fig. 11.5 obtained from the gas-phase optimized geometries in both the ground and excited states. The curve shows significant deviations from the scattering signal from QM/MM BOMD distributions. The differences are due to the gas-phase geometry optimization in the excited state predicting much larger changes in the structure of the ligands than those obtained from the calculations in solution.

In conclusion, the set of molecular structures used to model the solute difference scattering signal ($\Delta S^{\text{solu}}(q, t_\text{p})$) in the PtPOP XDS data, was obtained by varying the Pt–Pt distance of the ground-state geometry of the complex optimized in vacuum with GPAW, from 2.700 to 3.300 Å in steps of 0.001 Å, while keeping all the other atoms fixed. In this modelling framework, photoinduced structural changes in the structure of the solute are parametrized through only the Pt–Pt distance. Possible contributions to the signal arising from intramolecular changes other than the Pt–Pt bond contraction were found within the uncertainties of the measured signal, as described in Ref. [5]; a further confirmation that, in solution, structural changes affecting the ligands of PtPOP after photoexcitation are neglegible as compared to the Pt–Pt contraction.

11.3.2 Determining the Solute-Solvent Term

Figure 11.6 (Left) shows the isotropic difference scattering signal of PtPOP in water at a pump-probe time delay of 4.5 ps fitted with a model that includes only the $\Delta S_0^{\text{solu}}(q)$ and $\Delta S_0^{\text{solv}}(q)$ terms. $\Delta S_0^{\text{solu}}(q)$ used in the fit was obtained from the set of structures generated as explained in the previous paragraph. The fit has been performed by Postdoc E. Biasin among our experimental collaborators. The residuals of such a fit are usually interpreted, to a first approximation, as the signature of changes in solute-solvent distances that are not accounted for by only including $\Delta S_0^{\text{solu}}(q)$ and $\Delta S_0^{\text{solv}}(q)$ [16].

To substantiate the latter hypothesis we have computed $\Delta S_{\text{MD}}^{\text{solu−solv}}(q)$ according to Eq. (8.8) using the solute-solvent RDFs obtained from the equilibrated QM/MM trajectories for S_0 and S_1. The result is shown in Fig. 11.6 (Right), together with the

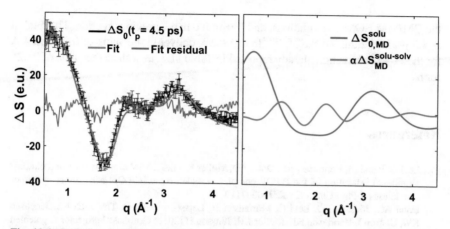

Fig. 11.6 (Left) Isotropic difference scattering signal of PtPOP in water at a pump-probe time delay of 4.5 ps fitted with a model including only the solute and solvent terms. (Right) Isotropic difference scattering signal simulated using the solute-solute (blue line) and solute-solvent (magenta line) equilibrium QM/MM RDFs for S_0 and S_1. The simulated solute-solvent term has been scaled by the experimental excitation fraction of 0.026. Differences between the fit to the experimental data and the simulated $\Delta S_{0,MD}^{solu}(q)$ arise mainly from the contribution due to heating of the bulk solvent

Fig. 11.7 Comparison between the residuals of the fit of the isotropic difference XDS signal of PtPOP in water at 4.5 ps, shown in Fig. 11.6 (Left), and the solute-solvent difference scattering signal simulated from QM/MM BOMD equilibrium distributions

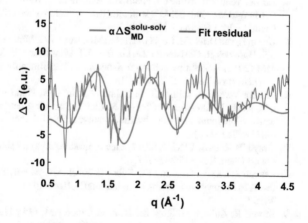

scattering from the solute ($\Delta S_{0,MD}^{solu}(q)$) calculated using the solute-solute RDFs, and already presented in Fig. 11.5; while Fig. 11.7 directly compares the residuals of the fit with $\Delta S_{MD}^{solu-solv}(q)$ scaled by the experimental excitation fraction α (α was equal to 0.026 and was obtained at a later stage of the analysis).

Indeed, the similarities between $\Delta S_{MD}^{solu-solv}(q)$ computed from the QM/MM BOMD data and the residual are remarkable. Once again, the interplay between simulations and experiments is mutually beneficial. First of all, a $\Delta S_0^{solu-solv}(q, t_p)$ term could be included in the fit of the time-dependent isotropic XDS signal, according to Eq. (8.19), thus improving the model. Secondly, the match between simulations and experiments corroborates the picture of (small) solvent shell changes drawn from the QM/MM BOMD simulations, and illustrated in the previous section. The final fit at 4.5 ps, including the $\Delta S_0^{solu-solv}(q, t_p)$ term, delivered the value of 0.24 ± 0.04 Å for the Pt–Pt contraction already reported in Table 11.1, as well as the value of 0.026 for α.

References

1. Levi G, Pápai M, Henriksen NE, Dohn AO, Møller KB (2018) Solution structure and ultrafast vibrational relaxation of the PtPOP complex revealed by ΔSCF-QM/MM direct dynamics simulations. J Phys Chem C 122:7100–7119
2. Dohn AO, Jónsson EÖ, Levi G, Mortensen JJ, Lopez-Acevedo O, Thygesen KS, Jacobsen KW, Ulstrup J, Henriksen NE, Møller KB, Jónsson H (2017) Grid-based projector augmented wave (GPAW) implementation of quantum mechanics/molecular mechanics (QM/MM) electrostatic embedding and application to a solvated diplatinum complex. J Chem Theory Comput 13(12):6010–6022
3. Christensen M, Haldrup K, Bechgaard K, Feidenhans R, Kong Q, Cammarata M, Lo Russo M, Wulff M, Harrit N, Nielsen MM (2008) Time-resolved X-ray scattering of an electronically excited state in solution. Structure of the a state of Tetrakis-μ-pyrophosphitodiplatinate (II). J Am Chem Soc 131(Ii):502–508
4. van der Veen RM, Milne CJ, Pham V-T, El Nahhas A, Weinstein JA, Best J, Borca CN, Bressler C, Chergui M (2008) EXAFS structural determination of the Pt2(P2O5H2)44- anion in solution. CHIMIA Int J Chem 62:287–290
5. Biasin E, van Driel TB, Levi G, Laursen MG, Dohn AO, Moltke A, Vester P, Hansen FBK, Kjaer KS, Hartsock R, Christensen M, Gaffney KJ, Henriksen NE, Møller KB, Haldrup K, Nielsen MM (2018) Anisotropy enhanced X-ray scattering from solvated transition metal complexes. J Synchrotron Radiat 25(2):306–315
6. van der Veen RM, Milne CJ, El Nahhas A, Lima FA, Pham VT, Best J, Weinstein JA, Borca CN, Abela R, Bressler C, Chergui M (2009) Structural determination of a photochemically active diplatinum molecule by time-resolved EXAFS spectroscopy. Angew Chem Int Edn 48(15):2711–2714
7. Gray HB, Záliš S, Vlček A (2017) Electronic structures and photophysics of d8–d8 complexes. Coord Chem Rev 345:297–317
8. Durrell AC, Keller GE, Lam YC, Sýkora J, Vlček A, Gray HB (2012) Structural control of 1A2u-to-3A2u intersystem crossing in diplatinum(II, II) complexes. J Am Chem Soc 134(34):14201–14207
9. Monni R, Auböck G, Kinschel D, Aziz-Lange KM, Gray HB, Vlček A, Chergui M (2017) Conservation of vibrational coherence in ultrafast electronic relaxation: the case of diplatinum complexes in solution. Chem Phys Lett 683:112–120
10. Záliš S, Lam Y-C, Gray HB, Vlček A (2015) Spin-orbit TDDFT electronic structure of diplatinum(II, II) complexes. Inorg Chem 54:3491–3500
11. Morse PM (1929) Diatomic molecules according to the wave Mechanics. II. Vibrational levels. Phys Rev 34:57–64
12. Schmidt B, Lorenz U (2017) WavePacket: a Matlab package for numerical quantum dynamics. I: closed quantum systems and discrete variable representations. Comput Phys Commun 213:223–234
13. Huber KP, Herzberg G (1979) Molecular spectra and molecular structure, vol IV. Constants of diatomic molecules. Van Nostrand Reinhold
14. van der Veen RM, Cannizzo A, van Mourik F, Vlček Jr A, Chergui M (2011) Vibrational relaxation and intersystem crossing of binuclear metal complexes in solution. J Am Chem Soc 113:305

15. Peterson JR, Kalyanasundaram K (1985) Energy- and electron-transfer processes of the lowest triplet excited state of Tetrakis(diphosphito)diplatinate(II). J Phys Chem 89(1983):2486–2492
16. van Driel TB, Kjær KS, Hartsock R, Dohn AO, Harlang T, Chollet M, Christensen M, Gawelda W, Henriksen NE, Kim JG, Haldrup K, Kim KH, Ihee H, Kim J, Lemke H, Sun Z, Sundstrom V, Zhang W, Zhu D, Møller KB, Nielsen MM, Gaffney KJ (2016) Atomistic characterization of the active-site solvation dynamics of a photocatalyst. Nat Commun 7:13678
17. Biasin E, van Driel TB, Kjær KS, Dohn AO, Christensen M, Harlang T, Chabera P, Liu Y, Uhlig J, Pápai M, Németh Z, Hartsock R, Liang W, Zhang J, Alonso-Mori R, Chollet M, Glownia JM, Nelson S, Sokaras D, Assefa TA, Britz A, Galler A, Gawelda W, Bressler C, Gaffney KJ, Lemke HT, Møller KB, Nielsen MM, Sundström V, Vankó G, Wärnmark K, Canton SE, Haldrup K (2016) Femtosecond x-ray scattering study of ultrafast photoinduced structural dynamics in solvated [Co(terpy)2]2+. Phys Rev Lett 117(1):013002
18. Dohn AO, Biasin E, Haldrup K, Nielsen MM, Henriksen NE, Møller KB (2015) On the calculation of x-ray scattering signals from pairwise radial distribution functions. J Phys B Atom Mol Opt Phys 48(24):244010. (Corrigendum: [19])
19. Dohn AO, Biasin E, Haldrup K, Nielsen MM, Henriksen NE, Møller KB (2016) Corrigendum: on the calculation of x-ray scattering signals from pairwise radial distribution functions. J Phys B Atom Mol Opt Phys 49(5):059501

Chapter 12
Coherent Vibrational Dynamics in the Ground State

In the present chapter, we scrutinize the set of S_1 nonequilibrium ΔSCF-QM/MM BOMD data and the S_0 nonequilibrium ground-state distributions of PtPOP in water established for a choice of pump-pulse parameters closely recreating the experimental conditions of the pump-probe X-ray diffuse scattering (XDS) measurements performed at LCLS. The details of the XDS experiments, realized during the present PhD project, can be found in Sect. 7.2; while the procedure for setting up the initial conditions for the dynamics has been described in Sect. 10.2.

Figure 12.1 shows the out-of-equilibrium distributions of Pt–Pt distances in S_0 and S_1 at time zero, and at times corresponding to half ($\sim$138 fs) and twice ($\sim$550 fs) the vibrational period of the ground state, respectively. The figure includes also the time-dependent hole distribution of Pt–Pt distances representing depletion of the ground-state ensemble, obtained as the (unnormalized) distribution of remaining ground-state molecules minus the ground-state equilibrium distribution. In Fig. 12.2 we show the full time evolution of the excited-state (blue density plot) and ground-state hole (red) distributions of Pt–Pt distances, together with the respective instantaneous Pt–Pt average distance (black curves).

Within the approximations used to set up the initial conditions (see Sect. 10.2), photoexcitation by the ultrashort pump laser creates a localized distribution in the excited state and a complementary localized hole in the ground-state equilibrium distribution. The nonequilibrium distributions, after time zero, start moving and spreading, equilibrating on the respective potential surfaces. Since the initial distributions are localized in out-of-equilibrium position, Pt–Pt distances within them start oscillating in phase, meaning that the vibrational motion is coherent. For the hole, this is a reflection of the remaining ground-state molecules vibrating coherently in the ground-state potential following d_{PtPt}-dependent depletion of the ground-state ensemble by the laser. Coherent vibrations occur with average periods of $\sim$276 and $\sim$227 fs for the ground and excited state, respectively. The period of $\sim$276 fs for S_0 is virtually identical to the one found from fitting the potential of mean force (PMF) of the equilibrium ground state distribution with a Morse potential (see Sect. 11.1),

© Springer Nature Switzerland AG 2019
G. Levi, *Photoinduced Molecular Dynamics in Solution*, Springer Theses,
https://doi.org/10.1007/978-3-030-28611-8_12

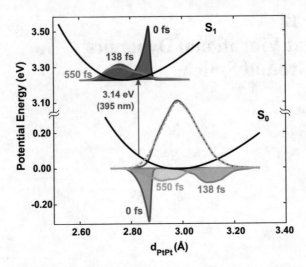

Fig. 12.1 Time-dependent S_0 and S_1 QM/MM distributions of Pt–Pt distances simulating dynamics following excitation by a laser with the parameters of the pump pulse used in the time-resolved XDS measurements on PtPOP in water. Shown are the distributions of the excited state (shaded blue areas), the remaining ground-state ensemble (red curves), and the ground-state hole (shaded red areas). For each state, the relative distributions are plotted at time zero, and at times after excitation corresponding to half and twice the vibrational period of the ground state. The excited-state and ground-state hole distributions, integrating to the simulated excitation fraction, are magnified for better visualization. All distributions were smoothed with a cubic smoothing spline

pointing, once again, to harmonicity of the Pt–Pt vibrations. The value obtained for the excited state is in very good agreement with the 224.5 ± 0.1 fs period observed by van der Veen et al. [2] in femtosecond optical measurements in water employing an excitation wavelength of 370 nm.

The photon energy of the excitation pulse ensures that the excited-state distribution is created very close to the equilibrium Pt–Pt distance in the S_1 state. Thus, the photoexcited molecules experience a small gradient along the Pt–Pt coordinate in the excited state, and the amplitude of the coherent vibrations are consequently small, as seen from Fig. 12.2(Top). The hole, on the other hand, starts from a position far from the equilibrium ground-state Pt–Pt distance and, therefore, undergoes large amplitude vibrations. This is the basis for the qualitative (and classical) understanding of the observation of only ground-state dynamics in the time-resolved XDS data of PtPOP in water presented in Sect. 7.2. In Fig. 12.3(Left), we have plotted the variation of the average Pt–Pt distance of the full simulated ensemble of PtPOP molecules computed from:

$$< d_{PtPt}(t) > = \int d_{PtPt} \left[P'_{ES}(d_{PtPt}, t) + P^{eq}_{GS}(d_{PtPt}) - P^{h}_{GS}(d_{PtPt}, t) \right] d(d_{PtPt})$$

$$= \int d_{PtPt} \left[P'_{ES}(d_{PtPt}, t) + P^{r}_{GS}(d_{PtPt}) \right] d(d_{PtPt}) \qquad (12.1)$$

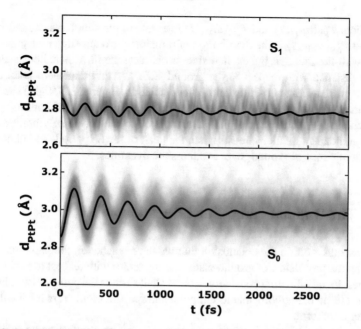

Fig. 12.2 Density plots of the time-dependent Pt–Pt distance distributions from the nonequilibrium ΔSCF-QM/MM trajectories in S_1 (Top) and S_0 nonequilibrium hole distributions (Bottom) obtained following photoexcitation of PtPOP in water by an ultrashort pulse selectively depleting the ground-state ensemble at short d_{PtPt}. The distributions were smoothed with a cubic smoothing spline. The superimposed black curves represent the respective instantaneous average Pt–Pt distances

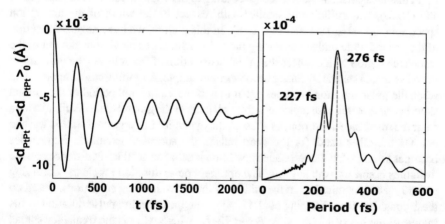

Fig. 12.3 Evolution of the Pt–Pt average distance computed from the ground- and excited-state time-dependent distributions of Pt–Pt distances (Left), and its FT (Right) showing peaks at the two vibrational periods characteristic of motion in S_0 (276 fs) and S_1 (227 fs)

where both $P^r_{ES}(d_{PtPt}, t)$ and $P^h_{GS}(d_{PtPt}, t)$ integrate to the simulated excitation fraction, and $P^r_{GS}(d_{PtPt}, t)$ is the distribution of remaining ground-state molecules. Consistent with the fact that the excited state is characterized by a smaller Pt–Pt equilibrium distance with respect to the ground state, we find that there is an overall decrease in the average Pt–Pt distance after excitation. Figure 12.3(Right) shows that the Fourier transform (FT) of $< d_{PtPt}(t) >$ delivers the periods of both the ground and excited states. The peak of the FT associated to the period of vibrations in S_1 has a significantly smaller intensity than the peak of the S_0 period, an indication of the predominance of ground- over excited-state dynamics.

12.1   The Optimal Pump-Pulse Duration

Let us expand a bit on the conclusion that we have reached in the previous section regarding the prevalence of ground-state hole dynamics with respect to excited-state dynamics. In the following considerations we shall assume the pump pulse is Fourier-transform limited, implying that its temporal and spectral profiles are related through a Fourier transform.

In the classical picture that we have adopted so far, an ultrafast probe can clearly detect dynamics in the ground state only if the excitation pulse creates a narrow, localized hole in position space. In our case, this was allowed by a long enough pulse, such that its frequency spread, determining the spread of the distribution of Pt–Pt distances that can be excited (according to the window function of Eqs. (10.4) and (10.5)), was sufficiently smaller with respect to the width of the absorption spectrum (see Table 10.1 and Fig. 10.2). In this picture, we have neglected motion of the ground-state molecules during the pulse. Molecules moving in and out of the resonance region would smear the initial distribution of the hole. As the duration of the pulse increases, the smearing becomes more ample. A limiting case is approached when the pulse duration becomes equal to half the vibrational period of the ground state, because in this case molecules from *all* parts of the ground-state distribution can enter the resonance condition; as a result, the dynamics in the ground state is washed out. On the other hand, for too short pulses, the spread of excitation frequencies becomes so large that the resonance condition is satisfied at all Pt–Pt distances, so the dynamics is smeared out anyway. The optimum pulse duration for clearly observing ground-state dynamics lies in the middle of these two limiting conditions. Based on these considerations, Fleming et al. [3, 4] define the optimum pulse duration as the middle of the interval $1/\Delta\nu_{abs} < \Delta\tau < T_{GS}/2$, where $\Delta\nu_{abs}$ is the frequency spread of the absorption spectrum and T_{GS} is the ground-state vibrational period. In the case of the pump pulse used in the XDS experiments on PtPOP, the duration of the pulse $\Delta\tau$, as quantified by the full width at half maximum (FWHM) of the temporal laser profile, is $\sim$50 fs (see Table 10.1), while the experimental $1/\Delta\nu_{abs}$ and $T_{GS}/2$ are, respectively, $\sim$20 fs (FWHM of the spectrum reported in Fig. 10.2) and $\sim$140 fs. Therefore, the pulse duration lies very close to the middle ($\sim$60 fs) of the interval

$1/\Delta\nu_{abs} < \Delta\tau < T_{GS}/2$. So, the experiments employed pump-pulse frequency and duration that, in this classical picture, are optimal for enhancing ground-state hole dynamics.

Obviously, the finite duration of the pulse has an effect also on the amplitude of the coherent oscillations in the excited state. The initial distribution will be broadened more, and hence the coherent dynamics will be smeared out faster, in the electronic state in which motion happens more rapidly. For PtPOP, the S_1 state has a higher Pt–Pt stretching frequency than the ground state. Therefore, we expect that the smearing due to the finite pulse duration is more significant for S_1 than for S_0. This could be a plausible explanation of why the experimental data contain no trace at all of excited-state dynamics (see FTs in Fig. (7.3)), while the simulations predict the presence of a (although small) contribution from dynamics in S_1.

12.2 Comparison with the Fit of the XDS Data

Figure 12.4 (Top) shows the results of the fit of the isotropic difference scattering signal from the PtPOP XDS data in terms of the best-fit Pt–Pt distance of the ground-state hole as a function of time. The fit has been performed by E. Biasin and K. Haldrup [1] employing the model presented in Sect. 8.2 (see especially Eq. (8.21) and related discussion), describing the hole with the set of single vacuum ground-state structures generated as explained in Sect. 11.3. The time-dependent d_{PtPt} could be fitted with an exponentially damped sine function multiplied with a step function centered at $t = 0$ and convoluted with a Gaussian with a width of $\sim$60 fs, representing the Instrument Response Function (IRF) of the experiment. The fitting function is also shown in Fig. 12.4(Top). The fit delivered a period of $\sim$284 fs and a decay time of the coherent oscillations of $\sim$1.7 ps. Figure 12.4(Bottom) shows the same type of fit as performed on the average Pt–Pt distance obtained from the time-dependent simulated hole distributions convoluted with the same IRF as employed in the fit of the experimental data. From this, we obtain a period of 271 fs, which agrees to within 5% with respect to the experimental period. The coherence decay from the simulations, on the other hand, is found to be around two times faster than from the experiments. We argue that the discrepancy could be a consequence of the simulations slightly overestimating the anharmonicity of the Pt–Pt vibrations. Indeed, the oscillation period of the simulated average d_{PtPt} hole distance changes by around 20 fs from the first oscillation to the oscillations at times longer than 1.5 ps; while no appreciable variation in the Pt–Pt vibrational period could be observed from the data. Another noticeable discrepancy is represented by a reduced amplitude of the first two oscillations in the Pt–Pt distance determined from the structural fit of the data with respect to the simulated d_{PtPt} hole distance. This has been interpreted as an indication of the presence of multi-photon excitation of PtPOP in the experiments.

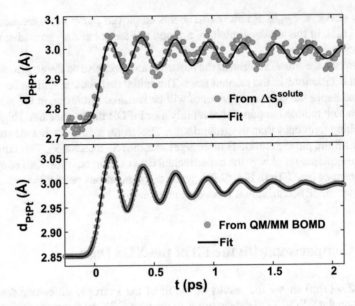

Fig. 12.4 Comparison between the time dependent Pt–Pt distance of the ground-state hole determined from the structural fit of the isotropic difference XDS signal (Top), and the same parameter obtained from the nonequilibrium QM/MM hole distributions convoluted with the IRF of the experiment (Bottom). Both experimental and simulated d_{PtPt} have been fitted with an exponentially damped sine function multiplied with a step function centered at $t = 0$ and convoluted with a Gaussian IRF (black continuous curves). The analysis of the experimental data has been performed by E. Biasin and K. Haldrup and is presented in Ref. [1]

12.3 Absorption or Raman?

It might appear surprising that, in discussing ground-state coherent vibrational dynamics induced by an ultrashort optical laser, we have not appealed to the so-called resonant impulsive stimulated Raman scattering (RISRS) process [3, 5–7]. RISRS is, usually, invoked to explain photoinduced vibrational dynamics in the ground state from a quantum mechanical viewpoint [2, 3, 8]. The quantum mechanical viewpoint is based on stationary vibrational eigenstates, and thus pictures ground-state dynamics in the energy space. The ground-state equilibrium ensemble is represented by a thermal density matrix defined as an incoherent sum of stationary eigenstates (see Eq. (11.2)). In this quantum mechanical picture, during interaction with an optical laser, two concomitant processes are in play: absorption and the RISRS process. Absorption transfers population from the thermally populated vibrational eigenstates of the ground state to the excited state. This event cannot explain coherent motion in the ground state, since the eigenstates do not acquire any time dependence after they have been depopulated, they remain stationary. Ground-state coherent oscillations must arise due to RISRS, from a quantum mechanical viewpoint. RISRS transfers population, during the pulse, from the excited state back to the ground state, thus leading to the formation of a coherent sum of stationary states, i.e. a wave packet. Such a wave packet is non-stationary and will move on the ground-state potential surface with the characteristic period of the ground state.

The classical picture of a moving hole localized in position space, originating from selective absorption of Pt–Pt distances, which we have used to interpret coherent ground-state dynamics, is only apparently incompatible with the picture offered by quantum mechanics. As explained by Fleming et al. in seminal works [3, 4], even though the classical picture has no authority to describe what happens during interaction with the laser pulse, it implicitly incorporates the combined effects of absorption and RISRS in the description of the resulting ground-state dynamics. These authors have shown that the vibrational dynamics in the ground state described using the two pictures can be surprisingly similar, especially at high temperature, when many vibrational eigenstates of the ground state are initially populated. In some limiting cases, the classical and the quantum mechanical viewpoints give essentially identical results, since they use two different ensembles with the same density matrix.

References

1. Haldrup K, Levi G, Biasin E, Vester P, Laursen MG, Beyer F, Kjær KS, Brandt van Driel T, Harlang T, Dohn AO, Hartsock RJ, Nelson S, Glownia JM, Lemke HT, Christensen M, Gaffney KJ, Henriksen NE, Møller KB, Nielsen MM (2019) Ultrafast x-ray scattering measurements of coherent structural dynamics on the ground-state potential energy surface of a diplatinum molecule. Phys Rev Lett 122:063001
2. van der Veen RM, Cannizzo A, van Mourik F, Vlček Jr A, Chergui M (2011) Vibrational relaxation and intersystem crossing of binuclear metal complexes in solution. J Am Chem Soc 113:305
3. Jonas DM, Bradforth SE, Passino SA, Fleming GR (1995) Femtosecond wavepacket spectroscopy: influence of temperature, wavelength, and pulse duration. J Phys Chem 99(9):2594–2608
4. Jonas DM, Fleming GR (1995) Vibrationally abrupt pulses in pump-probe spectroscopy. In: Ultrafast processes in chemistry and photobiology, chemistry for the 21st century, pp 225–256. Cambridge, Mass
5. Gershgoren E, Vala J, Kosloff R, Ruhman S (2001) Impulsive control of ground surface dynamics of I3- in solution. J Phys Chem A 105(21):5081–5095
6. Bartana A, Banin U, Ruhman S, Kosloff R (1994) Intensity effects on impulsive excitation of ground surface coherent vibrational motion. Chem Phys Lett 229:211–217
7. Bartana A, Banin U, Ruhman S, Kosloff R (1994) Impulsive excitation of coherent vibrational motion ground surface dynamics induced by intense short pulses. J Chem Phys 101(10):8461
8. Monni R, Auböck G, Kinschel D, Aziz-Lange KM, Gray HB, Vlček A, Chergui M (2017) Conservation of vibrational coherence in ultrafast electronic relaxation: the case of diplatinum complexes in solution. Chem Phys Lett 683:112–120

Chapter 13
Dynamics of Excited-State Bond Formation

In the present chapter, we describe the analysis of the second set of nonequilibrium ground-state distributions and ΔSCF-QM/MM trajectories of PtPOP in water. These were produced using the parameters of the pump pulse employed in the transient absorption setup that allowed van der Veen et al. [2] to probe the ultrafast vibrational dynamics upon excitation into S_1 of PtPOP in water. These excited-state simulations were primarily aimed at (i) characterizing the coherence decay of the Pt–Pt vibrations in S_1 in water, (ii) elucidating structural distortions involving the ligands during the excited-state dynamics and (iii) assessing whether the ultrafast relaxation is exclusively governed by specific solute-solvent interactions, as suggested in Ref. [2], or whether energy-accepting modes in the complex are also playing a role as mediators in the transfer of energy to the solvent, as hypothesized in Ref. [3]. To address the last point we have performed a vibrational analysis of the ΔSCF-QM/MM trajectories according to the method illustrated in Sect. 6.2.2 of the theoretical and computational methods part of the present thesis. The content of this chapter is included, with minor variations, in Ref. [1].

Due to the vast amount and statistical variability of parallel processes playing out in solution, extracting clear indications about the most likely paths of energy relaxation from a BOMD-generated out-of-equilibrium solution ensemble can be arduous if not impracticable at all. In fact, the interplay between anharmonic couplings and stochastic events can lead to incoherent processes, making the monitoring of average dynamical properties useless, while, at the same time, extrapolation of ensemble trends from the behaviour of a few individual uncorrelated trajectories can be dangerous, due to statistical bias. For this reason, we have performed additional ΔSCF-QM BOMD simulations of an isolated gas-phase PtPOP molecule in S_1 with the aim to gain preliminary insights into the excited-state intramolecular energy flow and, thus, facilitate the interpretation of the vibrational analysis of the

Parts of this chapter have been reproduced with permission from Ref. [1], https://doi.org/10.1021/acs.jpcc.8b00301. Copyright 2018 American Chemical Society.

© Springer Nature Switzerland AG 2019

G. Levi, *Photoinduced Molecular Dynamics in Solution*, Springer Theses, https://doi.org/10.1007/978-3-030-28611-8_13

165

S_1 ΔSCF-QM/MM trajectories. The ΔSCF-QM BOMD simulations in vacuum were performed by propagating the system with velocity Verlet with an integration time step of 1 fs. To allow a time step of 1 fs, all O-H bonds and hydrogen bonds present in PtPOP were constrained with the ASE implementation of RATTLE [4], as done for all QM/MM BOMD simulations performed in the present work. Likewise, the complex was described using GPAW with BLYP. During the dynamics, the translational and rotational degrees of freedom (DOF) were removed at each step by projecting out the total linear and angular momenta, respectively.

In the next section we explain how the vibrational velocities needed for the generalized normal mode analysis, illustrated in Sect. 6.2.2, were obtained from the solution-phase and vacuum trajectories. Section 13.2 is dedicated to presenting the results of the vacuum simulations; while Sect. 13.3 deals with the QM/MM BOMD simulations.

13.1   Extracting the Body-Fixed-Frame Velocities

Generalized normal modes and corresponding velocities were computed for the PtPOP complex from both gas-phase and solution-phase trajectories. To perform the analysis, the cartesian velocities without contributions from translation and overall rotation of the molecule are needed.

For the gas-phase simulations, since we have removed the total linear and angular momenta at each step of the propagation, the body-fixed-frame velocities were readily available. For the solution-phase trajectories, where it was not possible to separate out translation and overall rotation of the solute during the ΔSCF-QM/MM BOMD propagation, the body-fixed-frame velocities V_p to be used in (6.34) were obtained from the cartesian velocity vectors $\dot{\mathbf{R}}_a(t)$ by an a posteriori procedure. First of all, we required that the origin is at the molecule's center of mass, i.e. $\sum_a^{N_n} M_a \mathbf{R}_a(t) = 0$ and $\sum_a^{N_n} M_a \dot{\mathbf{R}}_a(t) = 0$, where N_n is the number of atoms in PtPOP, to separate the translation. Afterwards, we applied a rigid rotation to align all frames to a reference structure:

$$\mathbf{R}_a'(t) = \mathbf{A}(t)\mathbf{R}_a(t) \tag{13.1}$$

where the rotation matrix $\mathbf{A}(t)$ was computed using the Kabsch method [5], which minimizes the root-mean-square deviation (RMSD) between the instantaneous structure $\mathbf{R}'(t)$ and the reference frame. Finally, we assumed the overall rotational energy and internal kinetic energy of the molecule to be completely separable, such that the total kinetic energy is given by:

$$E_k(t) = \frac{1}{2}\dot{\mathbf{R}}_{CM}^2(t) \sum_{a=1}^{N_n} M_a + \frac{1}{2} \sum_{a=1}^{N_n} M_a \left(\boldsymbol{\omega}_r(t) \times \mathbf{R}_a'(t)\right)^2 + \frac{1}{2} \sum_{a=1}^{N_n} M_a \mathbf{V}_a^2(t) \tag{13.2}$$

where $\mathbf{R}_{CM}(t)$ is the translating position of the origin of the system of axis with respect to the fixed laboratory system and $\boldsymbol{\omega}_r(t)$ is the apparent angular velocity obtained from the instantaneous moment of inertia and angular momentum of the molecule ($\boldsymbol{\omega}_r(t) = \mathbf{I}^{-1}(t) \sum_a^{N_n} \mathbf{R}'_a(t) \times \dot{\mathbf{R}}'_a(t)$); and calculated the body-fixed-frame velocities according to:

$$\mathbf{V}_a(t) = \dot{\mathbf{R}}'_a(t) - \boldsymbol{\omega}_r(t) \times \mathbf{R}'_a(t) \tag{13.3}$$

When multiple trajectories were available, the average in Eq. (6.34) to compute the covariance matrix $\mathbf{K}$ was carried out over time and trajectories.

The Python script used to extract the body-fixed-frame velocities of PtPOP at each step of the ΔSCF-QM/MM trajectories is provided in Appendix A.

13.2 Gas-Phase Vibrational Dynamics

We have performed two different types of ΔSCF-QM BOMD simulations of S_1 PtPOP in vacuum. The initial gas-phase structures utilized in the two simulations are shown in Fig. 13.1.

In the first simulation (simulation (I)), a single S_1 trajectory was started from a structure optimized in vacuum in the S_1 state with respect to all DOF except for the Pt–Pt distance, which was set at the value of the ground-state structure optimized in vacuum using GPAW with BLYP (3.005 Å, see Sect. 9.3). At the beginning of the simulation, all atomic momenta were equal to 0. The trajectory was then propagated

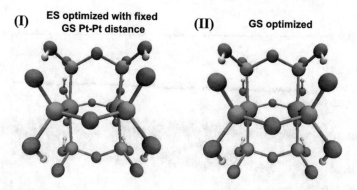

Fig. 13.1 The initial PtPOP structures used in the two ΔSCF-QM BOMD simulations in the S_1 excited state (ES). (Left) Simulation (I) was started from the gas-phase geometry optimized in S_1 with the Pt–Pt distance constrained to the Pt–Pt distance of the gas-phase ground-state (GS) optimized geometry of the complex (3.005 Å). (Right) Simulation (II) was started from the gas-phase ground-state optimized geometry. The structures are drawn with the Pt–Pt axis oriented horizontally to highlight that Pt_2P_4 groups in (I) and (II) are in a quasi-trigonal bipyramidal and square-based planar geometry, respectively

for 16 ps with time step of 1 fs. While this choice of initial conditions is far from being representative of the state created by excitation with an ultrashort laser, it nevertheless provides a useful means for more easily identifying vibrational modes of the molecule that couple more strongly to the Pt–Pt stretching mode, since at the beginning of the dynamics almost all excess potential energy will be concentrated in this mode.

In a second vacuum ΔSCF-QM BOMD simulation (simulation (II)), we have propagated a single S_1 trajectory starting from the optimized geometry of the ground state. With respect to the SF approximation used to set up initial conditions for the nonequilibrium ΔSCF-QM/MM BOMD simulations (see Sect. 10.2), this second choice of initial conditions corresponds to a CW (infinitely long) pump pulse, i.e. the excitation window of Eq. (10.5) is a delta function (this is the Bersohn-Zewail model, see for example Ref. [6]). This simulation was aimed at producing a picture of the dynamics that is closer to the events that take place in an ultrafast pump-probe experiment than the one that emerges from the first simulation. Total propagation time and time step were the same as those of the ΔSCF-QM BOMD run started from a relaxed S_1 geometry with the Pt–Pt distance of the ground state.

13.2.1 Analysis of the Energy Drift

Figures 13.2 and 13.3 analyse the energy drift in the two different vacuum ΔSCF-QM BOMD simulations of PtPOP in S_1.

Figure 13.2 refers to the run started from a geometry with the Pt–Pt distance of the vacuum ground-state optimized geometry and all other DOF relaxed in

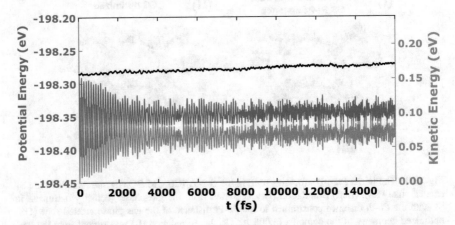

Fig. 13.2 Time dependence of the total potential and kinetic energies of PtPOP in the gas-phase ΔSCF-QM BOMD simulation (I) (see Fig. 13.1). The black curve represents the evolution of the total energy obtained as the sum of the potential and kinetic energies (uses the blue energy scale)

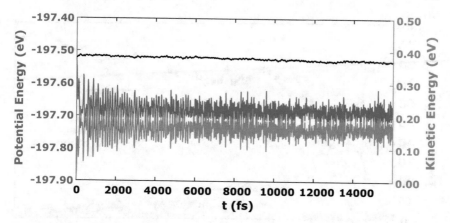

Fig. 13.3 Time dependence of the total potential and kinetic energies of PtPOP in the gas-phase ΔSCF-QM BOMD simulation (II) (see Fig. 13.1). The black curve is the instantaneous total energy (potential plus kinetic, uses the blue energy scale)

S_1 (simulation (I)). Figure 13.3 refers to the simulation started from the geometry optimized in vacuum in the ground state (simulation (II)).

In both cases, the total energy (potential plus kinetic) is satisfactorily stable throughout the entire simulation time of $\sim$16 ps. For the first simulation, at the end of the run the energy drift, quantified by the deviation of the instantaneous total energy from the value at $t = 0$, is equal to 1.3×10^{-2} eV (3.4×10^{-4} eV per atom), corresponding to a root-mean-square error (RMSE) of 7.7×10^{-3} eV. For the second simulation, the maximum energy drift in total energy is -0.9×10^{-2} eV (-2.4×10^{-4} eV per atom and RMSE of 4.7×10^{-3} eV).

13.2.2 Interplay Between Pinch, Twist and Breathing

We start by examining the trajectory from simulation (I) (see Fig. 13.1).

The Pt–Pt stretching mode takes up alone almost all excess vibrational energy at the beginning of the dynamics. This is apparent from Fig. 13.4, which reports at different intervals of time during the simulation the percent fraction of average total energy (kinetic plus potential) for the four modes that are found to have the largest average kinetic energy over the entire simulation time and for the sum of the rest.

The average total energy for each mode was calculated from the virial theorem as twice the average of the kinetic energy over intervals of 300 fs. The four selected modes are depicted in terms of generalized normal mode displacement vectors in Fig. 13.5. The figure also shows the Fourier transform (FT) of the autocorrelation function $C_p(t)$ of the mode velocities for each mode p. These were calculated from:

$$C_p(t) = \left\langle \dot{Q}_p(0)\dot{Q}_p(t) \right\rangle \tag{13.4}$$

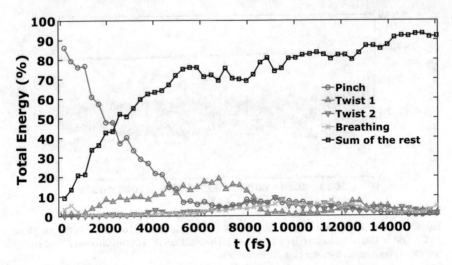

Fig. 13.4 Time evolution of the total energies (kinetic plus potential) of selected vibrational modes and the sum of the rest as obtained from a generalized normal mode analysis of the gas-phase S_1 trajectory from simulation (I). In total, the modes extracted from the vibrational analysis after removing the translations and overall rotations, and taking into account the constraints enforced on the positions of the hydrogen atoms during the dynamics, were 92. The total mode energies were averaged over time intervals of 300 fs and expressed as a percentage of the total average vibrational energy. See Fig. 13.5 for a depiction of the four selected modes

The positions of the FT peaks represent the characteristic vibrational frequencies of the modes.

The mode with character of Pt–Pt stretching is indicated as "pinch". Initially in the dynamics, this mode takes up to almost 90% of the total vibrational energy. After around 6 ps, the portion of energy shared by the pinching mode has decreased by around 95% of the initial value. Of this, ~80% has flowed into 87 modes, which seem to be activated simultaneously and at the same rate, with none of them showing particular preference for overtaking the excess Pt–Pt vibrational energy; while around 20% has been transferred to a single mode with main character of ligand twist (twist 1). Thereafter, a significant portion of the energy flow is directed towards the other two remaining modes (twist 2 and breathing), which, thus, seem to be activated rather sequentially after the activation of twist 1. The interplay between the four main modes identified in the vibrational analysis manifests itself in the evolution of the respective kinetic energies, as illustrated in Fig. 13.6. In particular, the strong coupling between the pinching mode and twist 1 is evident from the fact that while the kinetic energy of the twist 1 mode is at a maximum, around 6–7 ps, the kinetic energy of the pinch has reached a minimum, and after that has a small increase at the expense of the kinetic energy accumulated in twist 1. The same is true at early times in the dynamics for the pinching and breathing modes. In fact, at times earlier than 1 ps, kinetic energy is seen to rapidly flow in and out of the breathing mode, matching a local minimum in the evolution of the kinetic energy of the pinch.

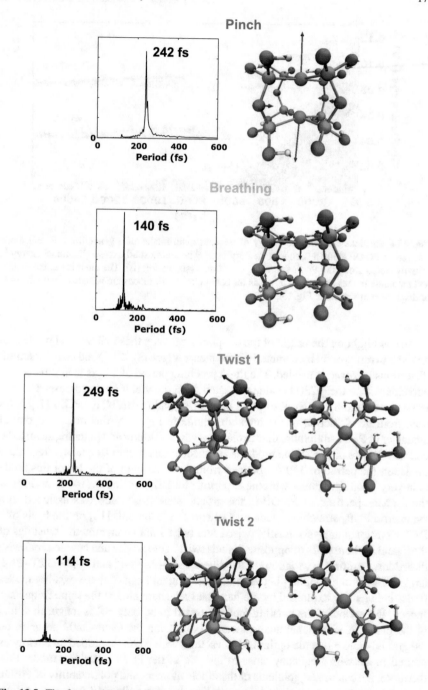

Fig. 13.5 The four main generalized normal modes involved in the gas-phase S_1 dynamics of PtPOP, and the FTs of their velocity autocorrelation functions. The position of the maximum of the FT of a mode gives the characteristic frequency of that mode

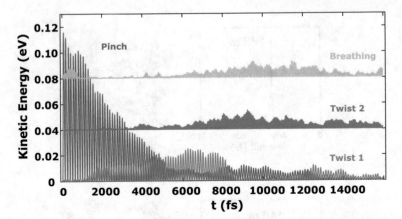

Fig. 13.6 Instantaneous kinetic energy of the four main modes of a generalized normal mode analysis of PtPOP along a vacuum trajectory in S_1 where almost all excess vibrational energy is initially stored along the Pt–Pt stretching coordinate (simulation (I)). The energies of the twist 2 and breathing modes are shifted upwards for better clarity. All modes are visualized with the help of displacement vectors in Fig. 13.5

To shed light on the origin of the couplings between these vibrational modes, an analysis in terms of their characteristic frequencies (see Fig. 13.5) and main structural distortions involved is needed. The Pt–Pt pinching period of 242 fs is in satisfactory agreement with the ~230 fs value extracted from the vibrational progression of the low-temperature $S_0 \rightarrow S_1$ absorption band of crystal (n-Bu$_4$N)$_4$[PtPOP] [7, 8]. As seen from the depiction of the breathing mode in Fig. 13.5, this mode has partial character of Pt–Pt stretching, thus explaining why breathing and pinching seem to be coupled despite the breathing mode has a considerably higher frequency. Regarding the latter, the period of 140 fs obtained from the maximum of the FT of this mode is in very good agreement with the experimental $232 \, \text{cm}^{-1}$ peak (144 fs period) of the Raman spectrum of PtPOP in the ground state [9], which was assigned to a symmetric Pt$_2$P$_8$ stretching mode by Gellene and Roundhill [10] on the basis of a DFT vibrational analysis. Lastly, twist 1 and twist 2 are antisymmetric twistings of the ligands, where pairs of opposite ligands twist in clockwise and counterclockwise directions, resulting in variations of the dihedral $\angle$P–Pt–Pt–P' and in-plane $\angle$P–Pt–P angles. The strong coupling between the pinching and one of these twisting modes (twist 1) is readily explained by the fact that they share almost the same frequency. Indeed, the period of the twisting mode is found to be only ~8 fs longer than that of the pinch. A vibrational analysis carried out using the Gaussian09 program on the ground-state molecule optimized in vacuum with BLYP identified an analogous normal mode with frequency close to the one of the Pt–Pt stretching mode. Furthermore, the calculated gradients of the dipole moment and polarizability of PtPOP along this mode are very close to 0, revealing that it is neither IR or Raman active, thus explaining why it has never been observed experimentally (unfortunately Gellene and Roundhill [10] have reported only DFT-calculated frequencies that could be

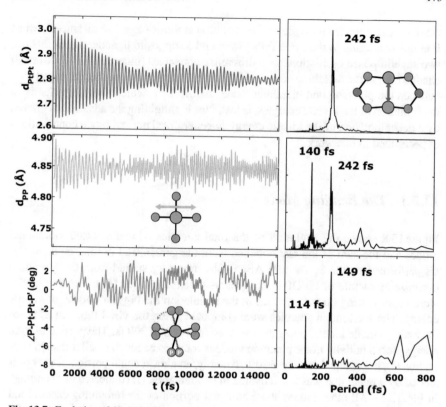

Fig. 13.7 Evolution of the main geometry parameters of PtPOP involved in the ΔSCF-QM BOMD vacuum simulation started from a relaxed S_1 structure with the Pt–Pt distance of the ground-state optimized geometry (simulation (I)), and their FT. Changes in the reported parameters and their frequencies correlate with variations in the kinetic energy of the normal modes presented in Fig. 13.6. (Top) Evolution of the Pt–Pt distance. (Middle) Instantaneous average P–P distance between P atoms belonging to opposite ligands. Since P–P symmetric vibrations take part in both the breathing and pinching modes, fluctuations in this parameter reflect the frequencies of both modes. (Bottom) Variation of the mean of the ∠P–Pt–Pt–P' dihedral angles involving ligands that undergo simultaneous clockwise torsion in the dynamics of the twist 1 and twist 2 modes

compared to experimentally determined IR or Raman transitions, thus a comparison with their vibrational analysis cannot be made for this mode).

Plots of the evolution of the structural parameters that are mostly involved in the dynamics of the selected modes, together with their FTs, are shown in Fig. 13.7. The frequencies of the fluctuations of reported atomic displacements and angles, and the time evolution of their amplitudes correlate very well with the evolution of the mode kinetic energies shown in Fig. 13.6, thus further validating the results of the generalized normal mode analysis. We note that it would be more difficult to infer the intramolecular energy flow from the local mode picture provided by the evolution of the amplitudes of oscillations of single structural parameters, when a priori knowledge of the coupled nuclear motion in the dynamics is lacking. This is clear, for example, from the plot of the evolution of the average P–P distance between P atoms belonging to the same PtP$_4$ group and to opposite ligands, which features two superimposed oscillations with different frequencies (indeed, the P–P distances change both in the breathing and pinching modes); as a consequence, the correlation between the pinching and breathing modes before 1 ps, which shows up clearly in the evolution of the kinetic energies, is lost, thus highlighting the advantages offered by a decomposition of the kinetic energy in generalized normal mode contributions as performed in this work.

13.2.3 The Bending Mode

Figure 13.8 shows the evolution of the total energies of five selected vibrational modes and the sum of the rest obtained from the generalized normal mode analysis performed on the S$_1$ vacuum ΔSCF-QM trajectory started from the gas-phase optimized structure of PtPOP in the ground state (simulation (II)). The five modes were those sharing on average during the simulation the biggest portion of kinetic energy. The total mode energies were computed using the virial theorem as twice the average mode kinetic energies over time intervals of 300 fs. The portion of total energy stored initially in the pinching mode is found to be much smaller than in simulation (I), being equal to only ~30%. Notably, an almost equal portion of energy is shared by a mode that was not activated in in simulation (I) (indicated as "bending" in Fig. 13.8). All other modes share an equal portion of the remaining excess total energy. Between ~1 and ~7 ps the flow of energy activates the twist 1 and breath-

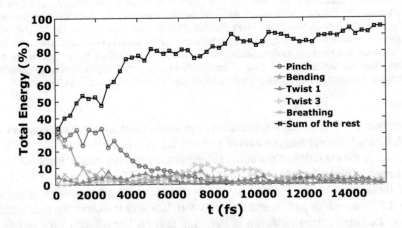

Fig. 13.8 Time evolution of the total energies (potential plus kinetic) of selected vibrational modes of PtPOP and the sum of the rest as obtained from a generalized normal mode analysis of the S$_1$ vacuum trajectory from simulation (II) (see Fig. 13.1). The total mode energies were averaged over time bins of 300 fs and expressed as a percentage of the total average vibrational energy of the molecule

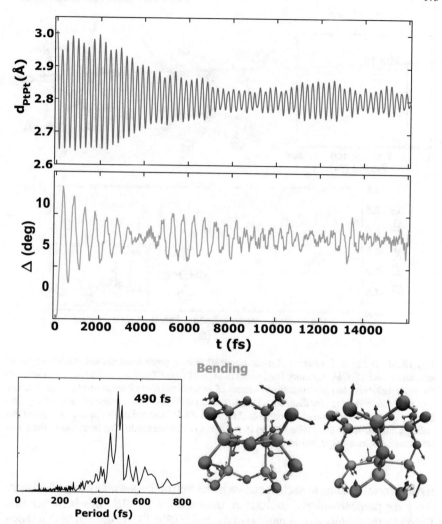

Fig. 13.9 Evolution of the vacuum S_1 trajectory started from the optimized gas-phase S_0 geometry of PtPOP (simulation (II)) along the coordinates that, at the beginning, share the largest portion of vibrational energy of the molecule. (Top) Variation of the Pt–Pt distance. (Middle) Evolution of the pseudorotation coordinate Δ defined in Fig. 9.8. (Bottom) Visualization of the mode that corresponds to motion along Δ and is activated at the beginning of the dynamics together with the pinching mode

ing modes (see Figs. 13.5 and 13.7 for a characterization of these modes), while at around 8 ps the mode indicated as "twist 3" has received a considerable portion of the excess total energy.

The bending mode is depicted in Fig. 13.9 (Bottom), where also the FT of the autocorrelation of mode velocities is reported. It is characterized by a bending of the

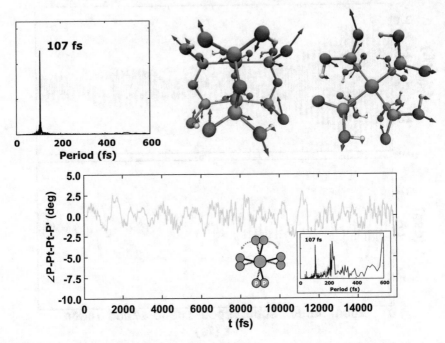

Fig. 13.10 The twist 3 vibrational mode obtained from a generalized normal mode analysis of the vacuum ΔSCF-QM trajectory from simulation (II). (Top) FT of the autocorrelation function of the mode velocities and representation in terms of generalized normal mode displacement vectors. The mode involves the simultaneous clockwise and counterclockwise twist of pairs of adjacent ligands. (Bottom) Variation of the mean of the $\angle$P–Pt–Pt–P' dihedral angles involving ligands that undergo simultaneous clockwise torsion in the dynamics of the mode. The inset shows the FT of the instantaneous average dihedral angle

ligands corresponding to nuclear motion in the well of the potential energy landscape along the pseudorotation coordinate Δ, as defined in Fig. 9.8. This is further confirmed by the evolution of Δ during the dynamics (Fig. 13.9 (Middle)), which shows a first rapid increase, followed by oscillations with a period of $\sim$490 fs around a value of about 7°, consistent with the shape and the minimum of the potential shown in Fig. 9.8. The only experimental indication of the existence of a vibrational mode with a lower frequency than the metal-metal stretching in PtPOP is given by the presence of a $\sim$40 cm^{-1} sideband on the Pt–Pt vibronic progression of low temperature absorption and emission spectra of single crystals of Ba$_2$[PtPOP] [11], which was attributed to a ligand deformation mode, but was never further characterized. According to the results of our simulations we assign the observed mode to a bending of the ligands in a D_{2d} geometry (of the Pt$_2$P$_8$ core), thus reaffirming the conclusion that PtPOP in the excited state does not retain a C_{4h} symmetry, which we have reached in Sect. 9.3 using PESs calculations.

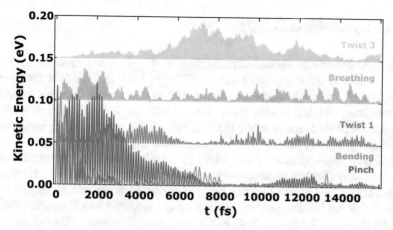

Fig. 13.11 Evolution of the instantaneous kinetic energy along the vibrational modes of PtPOP that undergo the largest displacements during the excited-state ΔSCF-QM BOMD simulation (I) started from the ground-state optimized geometry of the complex

The twist 3 mode is characterized in Fig. 13.10. Its behaviour is similar to the twist 2 mode observed for simulation (I), in that it is activated later in the dynamics, after about 6 ps, but has a slightly different period (∼107 fs) and character of the torsional motion.

In Fig. 13.11 we report the evolution of the instantaneous kinetic energy along the modes with the largest average kinetic energy over the entire BOMD simulation time. These include the pinching and bending modes, the breathing and twist 1 modes, already identified previously, and the twist 3 mode.

The kinetic energy along the bending mode is seen to decrease rapidly in the first ∼3 ps. The pinching mode, instead, has, at ∼3 ps, around the same kinetic energy it had at the beginning; hence, it is reasonable to assume that a considerable portion of the excess energy of the bending flows into the pinch. Besides, we observe how, in this second type of simulation, the coupling between the pinching and the breathing and twist 1 modes seems to be accentuated. This is apparent from the multiple local dips that characterize the evolution of the kinetic energy of the pinch in the first ∼3 ps, which are accompanied by variations of the same magnitude but opposite sign in the energies of the breathing and twist 1 modes. What is new with respect to simulation (I), in which all coordinates except the Pt–Pt distance were relaxed before running the dynamics, is that a non-negligible portion of the total kinetic energy of the molecule is also stored in the breathing and twisting modes from the very beginning of the dynamics. Therefore, we conclude that initial activation following the gradients of the excited-state potential after excitation can induce a stronger coupling of the ligand deformation modes with the Pt–Pt pinching mode, along the dynamics.

13.3 Dynamics in Solution

Figure 13.12 shows the excited-state distribution of Pt–Pt distances as obtained from the ensemble of out-of-equilibrium S_1 ΔSCF-QM/MM trajectories at three times: time zero, after around half the S_1 vibrational period, i.e. when the distribution is at the first inner classical turning point, and at a late time in the dynamics, when the coherent oscillations cease. While Fig. 13.13 shows density plots of the excited-state (blue) and ground-state hole (red) time-dependent d_{PtPt} distributions.

The laser pulse parameters used to set up the initial conditions for obtaining the non-stationary QM/MM ensembles create an initial distribution of Pt–Pt distances in S_1 that is localized within a narrow range of elongated distances with respect to the excited-state equilibrium position, while leaving a localized hole in the middle of the ground-state equilibrium distribution. The subsequent dynamics of the full ensemble is dominated by large amplitude coherent oscillations in S_1. Coherent Pt–Pt oscillations in the excited state are around the equilibrium distance of 2.79 Å, with a period of $\sim$230 fs, and persist up until around 2 ps. This vibrational period is slightly longer than the one ($\sim$227 fs, see previous chapter) obtained from the first set of ΔSCF-QM/MM trajectories, which were started closer to the bottom of the S_1 potential. Therefore, the potential appears characterized by a slight anharmonicity. No coherent vibrations can be seen from the plot of the center of the time-dependent ground-state hole distribution (Fig. 13.13 (Bottom)), as expected, but a periodic spreading and

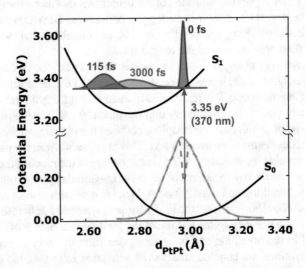

Fig. 13.12 Time-dependent excited-state distribution of Pt–Pt distances (blue shaded areas) as obtained from the nonequilibrium ΔSCF-QM/MM trajectories. The distribution is shown immediately after excitation, at its first inner turning point and at the end of the nonequilibrium dynamics. For the ground state, the figure shows the equilibrium distribution (grey line) and the portion of it remaining after excitation to S_1 (red dashed line) at time zero. The excited-state distributions are magnified for better clarity. All distributions were smoothed with a cubic smoothing spline

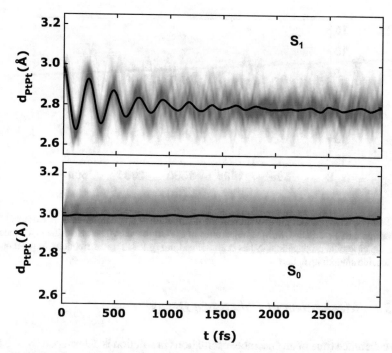

Fig. 13.13 Density plots of the evolution of the excited-state ensemble of Pt–Pt distances (Top) and of the ground-state hole (Bottom) obtained from QM/MM trajectories reflecting the initial conditions shown in Fig. 13.12. The distributions were smoothed with a cubic smoothing spline. The superimposed black curves represent the respective instantaneous average Pt–Pt distances

refocusing is apparent in the first $\sim$500 fs. Such a "breathing" of the hole distribution reflects the rigid rotation of the hole about the origin in the phase space [12–14] (Fig. 13.13).

In Sect. 9.3, we have shown that the Pt–Pt distance is not the only coordinate to undergo large changes from ground to excited state: the Pt–Pt contraction is accompanied by a bending of the ligands, quantified by an increase of the parameter Δ (see Fig. 9.8) by $\sim$5°. Therefore, we have examined the possibility that the ensemble of excited molecules displays coherent oscillations also along the coordinate Δ by plotting the evolution of the distribution of angle differences Δ (Fig. 13.14). As apparent from Fig. 13.14, no coherent oscillations are observed for the bending motion, but rather the trajectories along this coordinate exhibit the behaviour of overdamped oscillators, reaching the equilibrium value gradually over a time of $\sim$2.5 ps. This finding is consistent with the lack of oscillating signatures different from the Pt–Pt stretching vibrations in time-resolved measurements in solution [2].

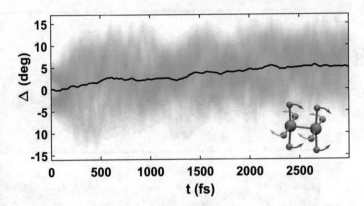

Fig. 13.14 Density plot of the evolution of the distribution of angle differences Δ extracted from the ensemble of S_1 ΔSCF-QM/MM trajectories of PtPOP in water. The black line is the mean Δ along the ensemble propagation. Δ has been defined in Fig. 9.8. The distributions were smoothed with a cubic smoothing spline

13.3.1 Mechanism of Coherence Decay

The coherence time of an ensemble oscillations in solution is determined by two processes happening concurrently: relaxation of vibrational energy, and pure dephasing events, the latter arising from elastic stochastic collisions with the solvent and phase changes along an anharmonic potential. An extensive discussion of the concepts of decoherence, vibrational cooling and pure dephasing in solution can be found in Ref. [15]. We have investigated the causes of decoherence of the Pt–Pt oscillations in the excited state by quantifying the time scales for coherence decay (τ_c), vibrational cooling (τ_e), and pure dephasing (τ_d) predicted by the simulations. To obtain the simulation decoherence time, we have fitted the time-dependent average of Pt–Pt distances with a periodic monoexponentially decaying function of the form (see Fig. 13.15).

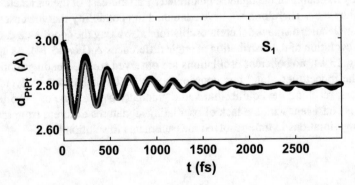

Fig. 13.15 Time dependence of the average Pt–Pt distance of the nonequilibrium S_1 ensemble (black line) together with the best fit (red line) from Eq. (13.5)

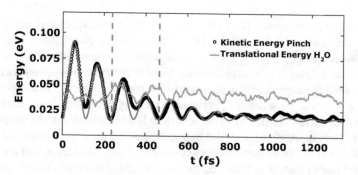

Fig. 13.16 Time dependence of the kinetic energy of the pinching mode of PtPOP (black open circles) obtained from the vibrational analysis of the S_1 ΔSCF-QM/MM BOMD simulations and averaged over all trajectories. The red line is the best fit of the function in Eq. (13.6) to the time dependence of the average pinching kinetic energy. Also shown (blue line) is the average translational kinetic energy of water molecules sampled by requiring that (i) at $t = 0$ they are within the first peak of the Pt-$O_{solvent}$ RDF (see Fig. 11.3) and (ii) at the end of the nonequilibrium propagation they have not left this coordination shell. Finally, the dashed vertical lines represent the times when the average Pt–Pt distance of the ensemble of non-stationary S_1 PtPOP molecules is at the first two outer turning points

$$f_c(t) = A e^{-t/\tau_c} \cos\left(\frac{2\pi}{T_{ES}}t\right) + B \tag{13.5}$$

in which T_{ES} is the coherent oscillation period. The vibrational cooling time τ_e was computed by fitting the time-dependent kinetic energy of the pinching mode obtained from the generalized normal mode analysis, and shown in Fig. 13.16, with the following function:

$$f_e(t) = C e^{-t/\tau_e} \cos^2\left(\frac{2\pi}{T_{ES}}t + \frac{\pi}{2}\right) + D \tag{13.6}$$

which reflects the dependence of the energy on the square of the relative velocities. The best fits gave values of $\tau_c = 520 \pm 14$ fs and $\tau_e = 320 \pm 10$ fs. The pure dephasing time τ_d was estimated by making use of the approximations underlying the optical Bloch equations [16]. In the optical Bloch picture, the rate of decoherence is given, phenomenologically, by the sum of the rates of vibrational cooling and pure dephasing:

$$\frac{1}{\tau_c} = \frac{1}{2\tau_e} + \frac{1}{\tau_d} \tag{13.7}$$

Using Eq. (13.7), a value of 2770 fs is found for τ_d. This means that the decoherence of the Pt–Pt vibrations is essentially driven by energy dissipation along the Pt–Pt coordinate, while statistical effects are far less important.

Experimentally, decoherence times of $\tau_c = 1.76 \pm 0.8$ ps and $\tau_c = 1.5 \pm 0.5$ ps were found from transient absorption andtime-resolved fluorescence up-conversion measurements respectively [2]. Furthermore, vibrational cooling was found to happen on time scales of $\tau_e = 1.31 \pm 0.04$ ps (transient absorption) and $\tau_e = 1.5 \pm 0.2$ ps (fluorescence up-conversion), i.e. simultaneously to coherence decay. Thus, while the coherence decay is around three times faster in our simulations, they agree qualitatively with the experiments in the observation that the origin of the decoherence is mostly dynamical, i.e. a result of (dynamical) energy dissipation in the excited system, and not statistical in nature. This behaviour is a consequence of the compactness and rigidity of the scaffold of P–O–P ligands, the first providing screening of the Pt–Pt oscillator from (stochastic) interactions with solvent molecules, and the second offering a highly harmonic force constant for the pinching motion (the period of the oscillations in the average Pt–Pt distance from the simulations changes by only ∼18 fs from the first to the last oscillation). As for the cause of the quantitative discrepancy between the coherence decay and vibrational relaxation times found for PtPOP from the present simulations with respect to the experimental values, we argue that this is a consequence of the calculations slightly overestimating the anharmonicity of the Pt–Pt motion. This is underpinned by the fact that the PMF computed from the equilibrated part of the S_1 ΔSCF-QM/MM trajectories, shown in Fig. 13.12, is best fitted with a Morse potential, while in the transient absorption measurements performed by van der Veen et al. [2] the period changes at most by ∼1.5 fs in going from a 360 to a 380 nm excitation wavelength (as already mentioned, in our simulations the period changes by ∼18 fs at the end of the coherent dynamics).

The mechanism of coherence decay in PtPOP is different from what was proposed [17] for the $[Ir_2(dimen)_4]^{2+}$ complex, already mention in Sect. 11.2. The main factor causing decoherence in $[Ir_2(dimen)_4]^{2+}$ is, in fact, statistical, ascribable to the flexibility of the dimen ligands that impart higher anharmonicity to the potential energy surface and a broader width to the distribution of configurations of ground-state molecules that can be excited [7, 17]. Even more insightful is, perhaps, a comparison with the behaviour observed for I_2 undergoing geminate recombination after photoexcitation in different environments. When the reaction was followed in solvents like CCl_4 or cyclohexane, vibrational relaxation was found to occur without coherent oscillations [18]. The behaviour of PtPOP is, instead, much more similar to that of I_2 in solid krypton, where stochastic collisions with solvent molecules leading to dephasing in solution are absent and the system is allowed to dissipate energy while retaining the vibrational phase [16, 19]. In all cases, the rigidity of the environment surrounding the oscillators is found to play an important role in determining whether vibrational coherence survives during the energy relaxation process or not.

13.3.2 Paths of Vibrational Energy Relaxation

Having established that vibrational cooling drives the coherence decay of the ensemble of Pt–Pt oscillators, the natural question that arises at this point is: what are the paths of energy dissipation from the Pt–Pt coordinate?

To provide an answer to this question, we have first investigated the hypothesis advanced by van der Veen et al. [2] that the main channel of energy dissipation involves transient orientationally specific interactions of the Pt atoms with water molecules coordinated at the open axial site. To do so, we have calculated the average translational and rotational energies of water molecules selected from the first solvent coordination shell around the Pt atoms defined by the first peak of the Pt-$O_{solvent}$ RDF, as indicated in Fig. 11.3. The average translational energy is plotted as a function of time in Fig. 13.16 together with the average kinetic energy along the pinching mode. The average rotational energy extracted from the trajectories did not display any particular displacement from its equilibrium thermal value. Early in the dynamics, the average translational energy of the coordinating water molecules experiences small positive fluctuations from its thermal equilibrium value. These fluctuations happen at around 250 and 450 fs, i.e. at the first and second outer turning points of the average Pt–Pt distance. This uptake of energy by the solvent, however, represents only a small fraction of the loss of energy from the pinching mode, and, certainly, cannot explain the steady decrease happening already during the first Pt–Pt oscillation period. In other words, the water molecules are more "spectators" of the Pt–Pt dynamics, rather than active participants in the relaxation process.

This is further substantiated by the time evolution of the first peak of the Pt-$O_{solvent}$ RDF presented in Fig. 13.17. The oscillations that appear until around 500 fs mirror the Pt–Pt oscillations of the excited-state ensemble of PtPOP molecules, thus implying that the solvent molecules are relatively static during this part of the dynamics. After that, the Pt–Pt distribution has almost reached an equilibrium, and the solvent molecules rearrange to the new solute configuration, as evident from the inset of Fig. 13.17, which reports the time evolution of the Pt-$O_{solvent}$ cumulative coordination number at the first minimum of the RDF ($d_{PtO} = 3.85$ Å). Since only

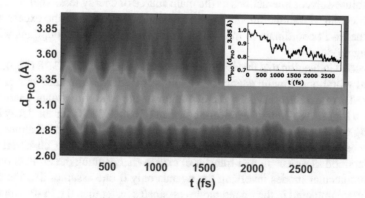

Fig. 13.17 Density plot of the time evolution of the first peak of the Pt-$O_{solvent}$ RDF obtained from the S_1 ΔSCF-QM/MM trajectories of PtPOP in water during the first 3 ps of dynamics. The inset shows the time dependence (black curve) of the cumulative Pt-$O_{solvent}$ coordination number at $d_{PtO} = 3.85$ Å, representing the instantaneous average number of water molecules found within the first solvent coordination shell of the Pt atoms, together with the value (red line) obtained from the equilibrated part of the ΔSCF-QM/MM trajectories

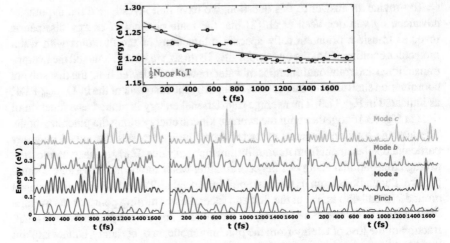

Fig. 13.18 (Top) Evolution of the ensemble average total vibrational kinetic energy of PtPOP (black circles) obtained as sum of the kinetic energies of the individual generalized normal modes, according to Eq. (6.36), from the vibrational analysis of the S_1 ΔSCF-QM/MM trajectories. The average ensemble total energy was further averaged over time intervals of 100 fs as indicated by the horizontal black lines. The red line is an exponential fit to the data points, while the horizontal dashed line represents the theoretical value of vibrational energy of an ensemble of molecules with the number of DOF of PtPOP in the simulations in equilibrium at 300 K. (Bottom) Plots of the time dependence of the kinetic energy along selected vibrational modes of PtPOP for three representative ΔSCF-QM/MM trajectories in S_1. The kinetic energies of modes a, b and c are vertically shifted for clarity of presentation

water molecules coordinating to the Pt atoms at the free axial sites are eligible to accept energy directly from the Pt–Pt pinching, the simulations seem to exclude direct solute-solvent interactions as the main source of energy loss. Therefore, other intramolecular vibrational modes have to mediate dissipation of the excess energy along the Pt–Pt coordinate to the solvent owing to anharmonic couplings with the pinching mode.

A second indication that this is indeed the case is given by Fig. 13.18 (Top), where the total vibrational kinetic energy of PtPOP averaged over time intervals of 100 fs is plotted. An exponential fit to the evolution of the total vibrational energy, also shown in Fig. 13.18 (Top), gives a time constant of 600 ± 200 fs for the decay before reaching equilibrium. Hence, the total vibrational energy is dissipated almost twice as slow as the vibrational cooling along the Pt–Pt coordinate. This observation can be understood as a clear sign of transfer of excess energy along the Pt–Pt coordinate to intramolecular modes involving the ligands only if one assumes that the energy initially accumulated in the ligand modes dissipates faster than the Pt–Pt vibrational cooling time. Since the ligands are exposed to direct interactions with the solvent it is reasonable to expect that the energy put into vibrational modes involving ligand atoms in the excitation process dissipates very efficiently to the solvent. It follows that the above result can be interpreted as an indication that the energy initially stored in the Pt–Pt coordinate survives in the PtPOP molecule for longer than the simulated vibrational cooling time for the ensemble of Pt–Pt oscillators.

As a last, more stringent, test of this mechanistic hypothesis we have plotted in Fig. 13.18 (Bottom) the evolution of the kinetic energy for the pinching mode together with three other relevant vibrational modes, as obtained from the generalized normal mode analysis, along three representative trajectories. The modes labelled mode b and mode c were found to have similar frequencies and large overlaps with respectively the twist 1 and breathing modes obtained from the gas-phase trajectories, and shown previously to be coupled to the pinching mode. However, they cannot be characterized fully as a twisting and a breathing mode, since they exhibit also character of other types of vibrations, most significantly Pt-P stretching (a representation of the modes in terms of displacement vectors is given in Fig. B.1 in Appendix B). Mode a does not overlap significantly with any of the main modes coupled in vacuum with the Pt–Pt pinch. It has mixed character of Pt-P stretching and ligand twist, with an autocorrelation function of mode velocities (see Fig. B.1) centered around 120 fs. This period is not far from the position of the peak (138 fs) in the IR spectrum of PtPOP assigned to P–Pt–P stretching by Gellene and Roundhill [10]. Notably, the evolution of the kinetic energies along these ligand deformation modes is seen from Fig. 13.18 (Bottom) to be strongly anticorrelated with the evolution of the kinetic energy of the pinching mode, since drops in the latter are always mirrored by increments of the former and vice versa. P-O and P-OH groups in the molecule experience large nuclear motion along the three ligand vibrational modes. Since these groups are likely involved in hydrogen bonding with the water molecules during the dynamics, the modes are expected to efficiently funnel excess energy to the solvent.

Overall, the simulations carry clear signs that dissipation of the Pt–Pt energy to the solvent, which drives the decoherence of the Pt–Pt oscillations, occurs mainly indirectly through IVR to modes characterized by motion of the O-P-OH moieties. The result seems to confirm the hypothesis recently put forward by Monni et al. [3], mentioned in the introduction, that anharmonic couplings between internal modes are the main source of decoherence of the Pt–Pt vibrations in photoexcited PtPOP. Since vibrational cooling along the pinching mode is found from the simulations to be much faster in solvent compared to vacuum, we can deduce that the role of the solvent is actually to facilitate anharmonic couplings between the modes, making IVR more efficient. Experimentally, van der Veen et al. [2] found a dependence of vibrational cooling on the solvent, which was interpreted as a signature of direct energy transfer from the Pt–Pt coordinate to the solvent. This interpretation, however, neglects the fact that different solvents can affect the strength of the anharmonic couplings between internal modes differently, thus changing the rates of IVR. Once again, this is in contrast to what was found for $[Ir_2(dimen)_4]^{2+}$ in acetonitrile, where the solvent prolongs coherence of the metal-metal oscillations by making, in some cases, IVR less likely than in vacuum [17].

References

1. Levi G, Pápai M, Henriksen NE, Dohn AO, Møller KB (2018) Solution structure and ultrafast vibrational relaxation of the PtPOP complex revealed by ΔSCF-QM/MM direct dynamics simulations. J Phys Chem C 122:7100–7119
2. van der Veen RM, Cannizzo A, van Mourik F, Vlček A Jr, Chergui M (2011) Vibrational relaxation and intersystem crossing of binuclear metal complexes in solution. J Am Chem Soc 113:305
3. Monni R, Auböck G, Kinschel D, Aziz-Lange KM, Gray HB, Vlček A, Chergui M (2017) Conservation of vibrational coherence in ultrafast electronic relaxation: the case of diplatinum complexes in solution. Chem Phys Lett 683:112–120
4. Andersen HC (1983) Rattle: a "velocity" version of the shake algorithm for molecular dynamics calculations. J Comput Phys 52:24
5. Kabsch W (1958) A solution for the best rotation to relate two sets of vectors. Acta Crystallogr Sect A 32:922–923
6. Petersen J, Henriksen NE, Møller KB (2012) Validity of the Bersohn-Zewail model beyond justification. Chem Phys Lett 539–540:234–238
7. Gray HB, Záliš S, Vlček A (2017) Electronic structures and photophysics of d8–d8 complexes. Coord Chem Rev 345:297–317
8. Stiegman AE, Rice SF, Gray HB, Miskowski VM (1987) Electronic spectroscopy of d^8-d^8 diplatinum complexes. $^1a_{2u}(d\sigma^* \to p\sigma)$, $^3e_u(d_{xz}, d_{yz} \to p\sigma)$, and $^{3,1}b_{2u}(d\sigma^* \to d_{x^2-y^2})$ excited states of $pt_2(p_2o_5h_2)_4^{4-}$. Inorg Chem 26:1112
9. Stein P, Dickson MK, Roundhill DM (1983) Raman and infrared spectra of binuclear platinum(II) and platinum(III) octaphosphite complexes. A characterization of the intermetallic bonding. J Am Chem Soc 105(12):3489–3494
10. Gellene GI, Roundhill DM (2002) Computational studies on the isomeric structures in the pyrophosphito bridged diplatinum (II) complex, platinum pop. J Phys Chem A 106:7617–7620
11. Rice SF, Gray HB (1983) Electronic absorption and emission spectra of binuclear platinum(II) complexes. Characterization of the lowest singlet and triplet excited states of $Pt_2(P_2O_5H_2)_4^{4-}$. J Am Chem Soc 105:4571–4575
12. Tannor DJ (2006) Introduction to quantum mechanics: a time dependent perspective. University Science Books
13. Jonas DM, Bradforth SE, Passino SA, Fleming GR (1995) Femtosecond wavepacket spectroscopy: influence of temperature, wavelength, and pulse duration. J Phys Chem 99(9):2594–2608
14. Jonas DM, Fleming GR (1995) Vibrationally abrupt pulses in pump-probe spectroscopy. In: Ultrafast processes in chemistry and photobiology, chemistry for the 21st century. Cambridge, Mass, pp 225–256
15. Kohen D, Tannor DJ (1997) Classical-quantum correspondence in the Redfield equation and its solutions. J Chem Phys 107(13):5141–5153
16. Jean JM, Fleming GR (2013) Competition between energy and phase relaxation in electronic curve crossing processes. J Chem Phys 103:2092–2101
17. Dohn AO, Jónsson EÖ, Kjær KS, van Driel TB, Nielsen MM, Jacobsen KW, Henriksen NE, Møller KB (2014) Direct dynamics studies of a binuclear metal complex in solution: the interplay between vibrational relaxation, coherence, and solvent effects. J Phys Chem Lett 5:2414–2418
18. Lee JH, Wulff M, Bratos S, Petersen J, Guerin L, Leicknam JC, Cammarata M, Kong Q, Kim J, Møller KB, Ihee H (2013) Filming the birth of molecules and accompanying solvent rearrangement. J Am Chem Soc 135:3255–3261
19. Zadoyan R, Li Z, Martens CC, Apkarian VA (1994) The breaking and remaking of a bond: caging of I2 in solid Kr. J Chem Phys 101(8):6648–6657

Part V
Concluding Remarks

Chapter 14
Summarizing Results and Outlook

The work presented herein focused on augmenting, benchmarking and applying a novel multiscale modelling strategy for simulating the structural dynamics of complex molecular systems in solution. The project has been prompted by the beginning of a new era of X-ray science, namely the "femtosecond era", in which modern sources of intense and ultrashort pulsed X-ray radiation enable the direct observation of the dynamics of the chemical bond in real time.

Femtosecond X-ray scattering measurements are emerging as a powerful tool to map photocatalytic processes in solution. Much of the attention is concentrated on elucidating the details of the ultrafast excited-state dynamics of transition metal complexes with photoconversion functionality. The interpretation and analysis of such novel ultrafast experiments call for first-hand theoretical support. Moreover, several experimental studies are starting to address the problem of improving efficiency and versatility of photocatalytic complexes by modifying their structure or by changing the solvent in which they are embedded [1–5]. Correctly linking the experimental outcomes to photocatalytic reactivity and tunability requires mechanistic knowledge, which, again, can only be attained through modelling and theory.

Assistance to the ultrafast experiments can be offered by atomistic simulations provided that they are both *reliable* and *efficient*. Detailed description of the solution dynamics of systems as large as transition metal complexes is far beyond the reach of multireference electronic structure methods, due to insurmountable computational requirements. On the other hand, entirely classical and empirical models cannot describe at ab initio level processes like bond-formation dynamics, coherence decay, energy transfer to the solvent. The route that we follow is based on the multiscale QM/MM coupling of a computationally expedient electronic structure code like GPAW with classical potential functions representing the solvent. The strategy allows for extensive sampling of solvent-influenced dynamics of a complex molecular system, within a Born-Oppenheimer Molecular Dynamics (BOMD) simulation framework.

© Springer Nature Switzerland AG 2019
G. Levi, *Photoinduced Molecular Dynamics in Solution*, Springer Theses,
https://doi.org/10.1007/978-3-030-28611-8_14

In the present work, we have expanded the capabilities of the original formulation of the GPAW-based QM/MM BOMD methodology, enabling it to describe electronic excited states with arbitrary spin multiplicity. In particular, we have chosen to try a cheap single-determinant DFT approach as ΔSCF. In Chap. 5 of this thesis we have provided the prerequisite theoretical background on GPAW and ΔSCF, and described the ΔSCF implementation. We have drawn upon already existing ΔSCF implementations that use a Gaussian smearing of the orbital occupation numbers to readily converge the electron density in a context of dynamically changing energy levels, and adapted the strategy within GPAW. The implementation has been tested on a diatomic molecule showing good reproducibility with respect to other, more standard, ΔSCF implementations, and further shown how a full potential energy surface can be reconstructed without convergence problems close to regions of states crossing thanks to the robustness provided by the Gaussian smearing.

A second part of the present project has dealt with the PtPOP molecule. PtPOP has received much attention in the last years as representative of a broader class of transition metal complexes with photocatalytic functionality. For this reason, it has been object of extensive ultrafast experimental studies. The model photocatalyst represented the ideal candidate for assessing the potentialities and performances of a combination of ΔSCF and QM/MM BOMD methodologies as applied to study the structural properties and dynamics of transition metal complexes. The assessment was in terms of both reproduction of previous experimental observations and assistance to new ultrafast X-ray scattering experiments with unprecedented time resolution carried out during this project.

In Chap. 9, we have reported the first calculated potential energy surfaces (PESs) of PtPOP along the Pt–Pt distance coordinate for both the lowest-lying singlet and triplet excited states, which provide the first computational evidence that they have approximately the same shape and position with respect to the ground-state gas-phase equilibrium geometry. While in Chap. 11, we have seen that the QM/MM BOMD simulations predict structural and dynamical properties in solution, such as the equilibrium Pt–Pt bond length, the excited-state structural changes and the Pt–Pt period of vibration, that are in agreement with experimental values. We have further elucidated the solvation shell structure in the ground and excited states, highlighting the presence of solvent molecules strongly coordinating along the Pt–Pt axis. The solvation cage is largely unaltered by excitation, a feature that, previously, had only been postulated based on experimental evidence. Ensemble properties have been robustly characterized using a large amount of statistics (around 460 ps for the ground state and more than 200 ps for the excited state), achieved thanks to the computational expediency inherent in GPAW and ΔSCF. We note that similar QM/MM studies on transition metal complexes [6–8] base their conclusions on statistical amounts of thermally sampled data which are roughly an order of magnitude smaller than those achieved in the present work.

We performed pump-probe X-ray diffuse scattering (XDS) experiments at an X-ray free electron laser (XFEL) on PtPOP in aqueous solution, where we followed the evolution of coherent Pt–Pt vibrations in the ground state. QM/MM BOMD simulations have been determinant in guiding the data analysis by refining the structural model (Chap. 11). Furthermore, they provided a semi-classical picture of the photoinduced dynamics of the full ensemble of molecules that is entirely useful in interpreting the experimental outcome. The picture is based on the formation of a localized hole in the ground-state distribution of Pt–Pt distances following laser excitation to the ground state. The model predicts the optimal experimental conditions for preparing a vibrationally "cold" excited-state population and a complementary narrow ground-state hole displaced from equilibrium, to bring out the vibrational modulation of the signal due to coherent ground-state dynamics. This was illustrated in Chap. 12.

Next, in Chap. 13, we took a step forward from the interpretation and validation of the experiments, and tried to uncover mechanistic aspects of the excited-state dynamics of PtPOP that had remained so far poorly understood, because they are not accessible by experiments. We summarize the main conclusions that we have reached about the ultrafast vibrational relaxation following photoexcitation of PtPOP to the first singlet excited state (S_1) in water:

- The Pt_2P_8 core of the molecule does not retain the D_{4h} symmetry it has in the ground state, as commonly believed, but distorts towards D_{2d} symmetry, following pseudorotation of the ligands. An aspect that had gone unnoticed in previous DFT studies, but which could play a decisive role in determining the trends observed in the ISC rates of PtPOP and its derivatives in solution.
- Decoherence along the Pt–Pt coordinate occurs through vibrational cooling while preserving to a large extent the vibrational phase.
- Channels of intramolecular vibrational energy redistribution (IVR) prevail over direct transfer to the solvent in determining the flow of excess Pt–Pt vibrational energy.
- The modes involved in the IVR have main character of ligand twisting and Pt–P bond stretching, and vibrational periods close to the period of the Pt–Pt stretching vibrations.
- The role of the solvent in the relaxation process is to strengthen anharmonic couplings between the pinching and the ligand deformation modes, thus facilitating IVR with respect to the scenario in vacuum.

Overall, cost-effective ΔSCF-QM/MM BOMD simulations appear, from the present study, as a powerful tool to investigate aspects of the excited-state dynamics and reactivity of complex molecular systems in solution. Therefore, they can be used to assist the analysis of, and complement ultrafast experiments for cases in which the BO approximation can be safely employed.

In this study, we have focused on the relaxation events taking place in S_1 in the first picoseconds after photoexcitation in water. Intersystem crossing to the lower lying T_1 state is known from transient absorption measurements to occur much later, after around $\sim$14 ps [9]. This permitted us to use ΔSCF-QM/MM BOMD simulations that neglect any non-adiabatic and spin-orbit couplings between electronic states. In our simulations, the S_1 state, where the investigated structural dynamics occurs, was found to be relatively well isolated from T_1 and other higher lying electronic states, as implied by the unperturbed shape of all obtained S_1 potential curves. However, fluctuations in the solvent configurations could temporarily shift the energy levels, thus favouring other electronic states getting closer to S_1. In order to asses the interplay between these transient energy levels fluctuations and the structural distortion of the symmetry of the molecule caused by pseudorotation of the ligands in determining the rates of ISC in water solution, non-adiabatic dynamics simulations including spin-orbit couplings (SOCs) and solvent effects are needed. Future computational studies should point in this direction to expand on the knowledge about the excited-state relaxation cascade at later times than those considered here.

In addition, the early events in the excited-state relaxation cascade in many photocatalytic transition metal complexes are dominated by couplings between electronic and nuclear motions. Electronic transitions can occur on picosecond or sub-picosecond time scales, and can play an important role in determining the catalytic properties of a complex. Therefore, many ultrafast experimental studies address the problem of determining the time scales of non-adiabatic processes in photoexcited transition metal complexes [10]. Recently, the femtosecond time resolution offered by XFELs was exploited to characterize coherent nuclear dynamics along with changes in the electronic character in a prototypical iron complex [11].

Hence, the next natural step to take to improve on the range of applicability of our code is to go beyond the BO approximation through inclusion of non-adiabatic effects in the dynamics. As mentioned in the introduction, this can be done, without losing the advantages offered by a classical description of the nuclear dynamics, using mixed quantum-classical schemes like the trajectory surface hopping (TSH) method. The development of surface hopping routines within ASE or the coupling of the GPAW ΔSCF implementation presented in this thesis with already integrated TSH programs, like SHARC [12, 13], represent possible projects for the future. The single-determinant character of ΔSCF, combined with the use of smooth orbitals, should guarantee efficient evaluation of the non-adiabatic coupling vectors, needed for the surface hopping propagation, using convenient finite difference methods [14, 15]. On the other hand, the approximation of utilizing a single-determinant method as ΔSCF to describe a problem that is inherently multiconfigurational will have to be rigorously assessed. Tests of the quality of non-adiabatic couplings computed with ΔSCF should be performed on small molecules for which high-level multireference methods are available, and on transition metal complexes against couplings calculated at TDDFT level, which is the current method of choice in non-adiabatic MD simulations of such systems.

References

1. Chábera P, Liu Y, Prakash O, Thyrhaug E, El Nahhas A, Honarfar A, Essén S, Fredin LA, Harlang TCB, Kjær KS, Handrup K, Ericson F, Tatsuno H, Morgan K, Schnadt J, Häggström L, Ericsson T, Sobkowiak A, Lidin S, Huang P, Styring S, Uhlig J, Bendix J, Lomoth R, Sundström V, Persson P, Wärnmark K (2017) A low-spin Fe(III) complex with 100-ps ligand-

to-metal charge transfer photoluminescence. Nature 543(7647):695–699

2. Liu L, Duchanois T, Etienne T, Monari A, Beley M, Assfeld X, Haacke S, Gros PC (2016) A new record excited state 3MLCT lifetime for metalorganic iron(II) complexes. Phys Chem Chem Phys 18:12550–12556

3. Harlang TCB, Liu Y, Gordivska O, Fredin LA, Ponseca CS, Huang P, Chábera P, Kjaer KS, Mateos H, Uhlig J, Lomoth R, Wallenberg R, Styring S, Persson P, Sundström V, Wärnmark K (2015) Iron sensitizer converts light to electrons with 92% yield. Nat Chem 7(11):883–889

4. Liu Y, Harlang T, Canton SE, Chábera P, Suarez-Alcantara K, Fleckhaus A, Vithanage DA, Goransson E, Corani A, Lomoth R, Sundström V, Warnmark K (2013) Towards longer-lived metal-to-ligand charge transfer states of iron(II) complexes: an N-heterocyclic carbene approach. Chem Commun 49(57):6412–6414

5. El Nahhas A, Cannizzo A, Van Mourik F, Blanco-Rodríguez AM, Záliš S, Vlček A, Chergui M (2010) Ultrafast excited-state dynamics of [Re(L)(CO)3(bpy)]n complexes: involvement of the solvent. J Phys Chem A 114(22):6361–6369

6. Penfold TJ, Curchod BFE, Tavernelli I, Abela R, Rothlisberger U, Chergui M (2012) Simulations of x-ray absorption spectra: the effect of the solvent. Phys Chem Chem Phys 14:9444

7. Daku LML, Hauser A (2010) Ab initio molecular dynamics study of an aqueous solution of [Fe(bpy)$_3$](Cl)$_2$ in the low-spin and in the high-spin states. J Phys Chem Lett 1:1830–1835

8. Moret M-E, Tavernelli I, Rothlisberger U (2009) Combined QM/MM and classical molecular dynamics study of [Ru(bpy)3]2+ in Water, 113:7737–7744

9. van der Veen RM, Cannizzo A, van Mourik F, Vlček Jr A, Chergui M (2011) Vibrational relaxation and intersystem crossing of binuclear metal complexes in solution. J Am Chem Soc 113:305

10. Chergui M (2015) Ultrafast photophysics of transition metal complexes. Acc Chem Res 48:801–808

11. Lemke HT, Kjær KS, Hartsock R, Van Driel TB, Chollet M, Glownia JM, Song S, Zhu D, Pace E, Matar SF, Nielsen MM, Benfatto M, Gaffney KJ, Collet E, Cammarata M (2017) Coherent structural trapping through wave packet dispersion during photoinduced spin state switching. Nat Commun 8(May):15342

12. Mai S, Marquetand P, González L (2015) A general method to describe intersystem crossing dynamics in trajectory surface hopping. Int J Quantum Chem 115:1215–1231

13. Richter M, Marquetand P, González-Vázquez J, Sola I, González L (2011) SHARC: ab initio molecular dynamics with surface hopping in the adiabatic representation including arbitrary couplings. J Chem Theory Comput 7(5):1253–1258

14. Plasser F, Ruckenbauer M, Mai S, Oppel M, Marquetand P, González L (2016) Efficient and flexible computation of many-electron wave function overlaps. J Chem Theor Comput 12(3):1207–1219

15. Hammes-Schiffer S, Tully JC (1994) Proton transfer in solution: molecular dynamics with quantum transitions. J Chem Phys 101(6):4657

Appendix A
Codes

Below we include scripts and parts of larger codes that have been developed in the course of the present project. The first entry contains the most important part of the Gaussian smearing ΔSCF code implemented in GPAW, and currently available only within a development branch of the program on Gitlab (https://gitlab.com/glevi/gpaw/tree/Dscf_gauss). We include it here, together with an example script for a Gaussian smearing ΔSCF calculation in GPAW, in the hope that it can serve as guidance in case someone intends to use the implementation, or wants to contribute to further develop it. We also provide scripts for extracting body-fixed frame cartesian velocities from MD trajectories and for performing a generalized normal mode analysis. We think these scripts might turn useful to students that are confronted with similar problems, or be source of inspiration for development within simulation packages like ASE.

Listing A.1 Python class developed in the GPAW module `occupations.py` for determining Gaussian smeared ΔSCF constraints on the orbital occupation numbers during an SCF cycle of a GPAW calculation. This is the most important part of the Gaussian smearing ΔSCF implementation. The implementation is currently available within the following development branch of GPAW: https://gitlab.com/glevi/gpaw/tree/Dscf_gauss. Projects to merge the implementation in the official release of the program are ongoing.

```python
class FixedOccupations_Gauss(ZeroKelvin):
    def __init__(self, occupation, constraints, width=0.01):
        self.occupation = np.array(occupation)
        self.constraints = constraints
        ZeroKelvin.__init__(self, True)
        self.width = width/Hartree
        self.niter = -1

    def spin_paired(self, wfs):
        return self.fixed_moment(wfs)

    def fixed_moment(self, wfs):
        for kpt in wfs.kpt_u:
            new_occupation = self.distribute_gaussian(kpt, self.
                occupation[kpt.s])
            wfs.bd.distribute(new_occupation, kpt.f_n)
```

© Springer Nature Switzerland AG 2019
G. Levi, *Photoinduced Molecular Dynamics in Solution*, Springer Theses,
https://doi.org/10.1007/978-3-030-28611-8

```
        # Fix the magnetic moment for spin polarized calculations
        if self.occupation.shape[0] == 2:
            self.magmom = self.occupation[0].sum() - self.occupation
                [1].sum()
            if self.constraints[0]:
                for orb in self.constraints[0]:
                    self.magmom += orb[0]
            if self.constraints[1]:
                for orb in self.constraints[1]:
                    self.magmom -= orb[0]

    def distribute_gaussian(self, kpt, ThisSpin_occupation):

        new_occupation = ThisSpin_occupation

        if self.constraints[kpt.s]:
            for c, orb in enumerate(self.constraints[kpt.s]):
                dx2 = (kpt.eps_n-kpt.eps_n[orb[1]])**2
                fgauss = 1/(self.width*np.sqrt(2*np.pi))*np.exp(-dx2
                    /(2*self.width**2))
                if orb[0]<0:
                    fgauss[self.occupation[kpt.s]==0]=0
                else:
                    fgauss[self.occupation[kpt.s]!=0]=0
                fgauss /= sum(fgauss)
                # Normalize the gaussian distribution such that
                # the sum of the smeared constraints is
                fgauss *= orb[0]
                # The constraints can be < or > 0
                # < 0 electrons are removed
                # > 0 electrons are added
                new_occupation = new_occupation+fgauss

        return new_occupation

    def todict(self):
        return {}
```

Listing A.2 Example of input script for a calculation with the Gaussian smearing ΔSCF implementation in GPAW. The script runs calculations to compute the energies of the first singlet and triplet excited states of the CO molecule.

```
from __future__ import print_function
from ase.parallel import paropen
from ase.structure import molecule
from gpaw import GPAW
from gpaw import Mixer, MixerSum, MixerDif
from gpaw.occupations import FixedOccupations_Gauss as FOG
import HPCPath as p
from gpaw.eigensolvers import CG
from gpaw.eigensolvers import Davidson
import os
from ase.io import read, write

PATH = p.HPCPath().path

#Define a name for the output files
fname = 'CO_lcao_0.18_tzpLDA_S2'
jobid=os.environ['PBS_JOBID']

#The energy of the optimized ground state
E_gs = -14.687650
```

```python
CO = read(PATH+'CO_lcao_0.18_tzpLDA_optSO.xyz')

#Set cell to cell of GS optimization
CO.set_cell([12, 12, 13.15034])

# Excited state calculation - Triplet
#---------------------------------------------------------
occupations=[[1, 1, 1, 1, 1, 0, 0, 0], [1, 1, 1, 1, 1, 0, 0, 0]]
calc_esT = GPAW(mode='lcao', basis='tzp', nbands=8, h=0.18, xc='LDA',
     spinpol=True,
                occupations=FOG(occupations, [[-1, 2]], [[1, 5]]], width
                    =0.01),
                maxiter_smear=80, maxiter=1000,
                convergence={'energy': 0.0005,
                             'density': 1.0e-4,
                             'eigenstates': 4.0e-8,
                             'bands': -1}, txt=PATH+fname+'_T2fromS0opt.
                                out')

CO.set_calculator(calc_esT)
E_esT = CO.get_potential_energy()
d=CO.get_distance(0,1)

# Excited state calculation - Singlet spin polarized
#---------------------------------------------------------
calc_esS_sp = GPAW(mode='lcao', basis='tzp', nbands=8, h=0.18, xc='
    LDA', spinpol=True,
                occupations=FOG(occupations, [[-1, 2],[1, 5]],[]], width
                    =0.01),
                maxiter_smear=80, maxiter=1000,
                convergence={'energy': 0.0005,
                             'density': 1.0e-4,
                             'eigenstates': 4.0e-8,
                             'bands': -1}, txt=PATH+fname+'
                                _S1fromS0opt_sp.out')

CO.set_calculator(calc_esS_sp)
E_esSsp = CO.get_potential_energy()

# Excited state calculation - Singlet spin paired
#---------------------------------------------------------
occupations=[[2, 2, 2, 2, 2, 0, 0, 0]]
calc_esS_ns = GPAW(mode='lcao', basis='tzp', nbands=8, h=0.18, xc='
    LDA', spinpol=False,
                occupations=FOG(occupations, [[-1, 2],[1, 5]]], width
                    =0.01),
                maxiter_smear=80, maxiter=1000,
                convergence={'energy': 0.0005,
                             'density': 1.0e-4,
                             'eigenstates': 4.0e-8,
                             'bands': -1}, txt=PATH+fname+'
                                _S1fromS0opt_ns.out')

CO.set_calculator(calc_esS_ns)
E_esSns = CO.get_potential_energy()

fd=paropen(PATH+fname+'.txt', 'w')
print(fd.name+'_____'+jobid, file=fd)
print(file=fd)
print('Basis_set:_____tzp', file=fd)
print('Goemetry_(Ang)___T_Potential_energy_(eV)', file=fd)
print('%.3f_____%.6f' % (d, E_esT), file=fd)
print('Excitation_energy_5sigma->2pi_T:___%.2f' %(E_esT-E_gs), file=
    fd)
print(file=fd)
print('Goemetry_(Ang)___S_sp_Potential_energy_(eV)', file=fd)
print('%.3f_____%.6f' % (d, E_esSsp), file=fd)
```

```
print('Excitation_energy_5sigma->2pi_S:___%.2f' %(2*E_esSsp-E_esT-
    E_gs), file=fd)
print(file=fd)
print('Goemetry_(Ang)___S_ns_Potential_energy_(eV)', file=fd)
print('%.3f_____%.6f' % (d, E_esSns), file=fd)
print('Excitation_energy_5sigma->2pi_S:___%.2f' %(E_esSns-E_gs), file
    =fd)
```

Listing A.3 Matlab script for performing a generalized normal mode analysis of an MD trajectory.
It takes as input a .dcd file with cartesian velocities.

```
%%%%%%%%%%%%%%%%%%%%%%%%%%%%%%%%%%%%%%%%%%%%%%%%%%%%%%%%%%%%%%%%%%%%%%%%%%%
% Calculate generalized Normal Modes from covariance of mass
% weighted cartesian velocities.
%
% Follows Strachan, A. JCP 120 (2004)
%
%                                                           G. Levi 2017
%%%%%%%%%%%%%%%%%%%%%%%%%%%%%%%%%%%%%%%%%%%%%%%%%%%%%%%%%%%%%%%%%%%%%%%%%%%

%% Define variables

kb = 8.617330337217213e-05;                             % eV/K
tstep = 1;                                              % fs
v2eV = 1e10*sqrt(1.6021766208e-19/1.660539040e-27); % 1e10*sqrt(eV2J*
    Na)
mass = [195.084,   195.084,    30.974,    30.974,    30.974,  ...
          30.974,    30.974,    30.974,    30.974,    30.974,  ...
          15.999,    15.999,    15.999,    15.999,    15.999,  ...
          15.999,    15.999,    15.999,    15.999,    15.999,  ...
          15.999,    15.999,    15.999,    15.999,    15.999,  ...
          15.999,    15.999,    15.999,    15.999,    15.999,  ...
           1.008,     1.008,     1.008,     1.008,     1.008,  ...
           1.008,     1.008,     1.008];
syms = {'Pt',  'Pt',  'P',  'P',  'P',  'P',  'P',  'P',  'P',  'P',  ...
        'O',   'O',   'O',  'O',  'O',  'O',  'O',  'O',  'O',  'O',  ...
        'O',   'O',   'O',  'O',  'O',  'O',  'O',  'O',  'O',  'O',  ...
        'H',   'H',   'H',  'H',  'H',  'H',  'H',  'H'};
mass = repmat(mass, 3, 1);
mass = reshape(mass, 1, 3*size(mass, 2));

%% Read in cartesian velocities

path = 'Path\to\trajectory\file\';
flname = 'TrajectoryFileName';                  % .dcd file with velocities

disp(['Reading_velocities_...'])

% Read in velocities
h = read_dcdheader([path flname]);       % Ang/ps
natoms = h.N;
nmols = 1;
nsteps = h.NSET;
startstep = 0;                                  % First step 0
step = 1;
laststep = nsteps-1;
nsteps_sel = ceil((laststep+1-startstep)/step);
t = 0:tstep:(nsteps_sel-1)*tstep;
vels = readdcd([path flname '_v.dcd'], startstep, step, laststep, 1:
    natoms);
vels = vels / (v2eV*1e-12);

%% Make covariance matrix of mass weighted velocities

% Mass weigthed velocities
```

```
mvels = repmat(mass.^(0.5), nsteps_sel, 1) .* vels;

disp(['Making_covariance_matrix_from_mass_weighted_velocities_...'])

% Covariance matrix
K = zeros(3*natoms, 3*natoms);
for aa = 1:3*natoms
    K(aa, aa:end) = 0.5 * mean(mvels(:, aa:end).*repmat(mvels(:, aa),
        1, 3*natoms-(aa-1)), 1);
    K(aa:end, aa) = K(aa, aa:3*natoms);
end

ekinm = trace(K);    % Average kinetic energy (eV)

%% Diagonalize covariance matrix of mass weighted velocities

disp(['Finding_normal_modes_...'])

% Now diagonalize covariance matrix
[L, em] = eig(K);
Lt = L';

%% Get NMs velocities

mVels = zeros(nsteps_sel, 3*natoms);
for tt = 1:nsteps_sel
    mVels(tt, :) = Lt * mvels(tt, :)';
end

% Calculate NMs kinetic energies
ekin = 0.5 * mVels.^2;

%% NMs total energies

t_wbin = 300;        % fs
t_edges = 0:t_wbin:(nsteps-1)*2;
[N, edges, bins] = histcounts(t, t_edges);
t_binned = zeros(1,length(N));
for ii = 1:length(N)
    t_binned(ii) = (t_edges(ii+1)+t_edges(ii))/2;
end
etot = zeros(length(N), 3*natoms);

% Calculate total energy
for nn = 1:3*natoms
    for ii = 1:length(N)
        etot(ii, nn) = 2 * mean(ekin(bins==ii, nn));
    end
end

%% Mode frequencies from FT of autocorrelation mode velocities

for nm =1:3*natoms
    disp(['Getting_autocorrelation_function_mode_' num2str(nm)])

    % Get autocorrelation function
    acV = zeros(nsteps_sel, 1);
    for tt = 1:nsteps_sel
        acV(tt) = mVels(1, 114-nm+1)*mVels(tt, 114-nm+1);
    end

    % Fourier Transform
    this_pad = 2^nextpow2(length(t));
    [T, Sft_spec] = dft_01(t, acV, this_pad);
    [M, idxM] = max(Sft_spec);
    fig = figure(1000)
    clf
```

```
     plot(T, Sft_spec, 'k', 'linewidth', 2)
     xlim([0 600])
     xlabel('Period_(fs)', 'interpreter', 'tex', 'fontsize',26);
     ylabel('a.u.', 'interpreter', 'tex', 'fontsize', 26);
     thisax = gca;
     textx = thisax.XLim(2) - (thisax.XLim(2)-thisax.XLim(1))*1/3;
     texty = thisax.YLim(2) - (thisax.YLim(2)-thisax.YLim(1))*1/3;
     text(textx, texty, {['T_=_' num2str(T(idxM), '%.1f') '_fs']}, '
          interpreter', 'tex', 'fontweight', 'bold', 'fontsize', 22)
     set(gcf, 'Position', [0 0 600 500])
     set(gca,'fontsize', 22,'fontweight', 'bold', 'LineWidth', 1.5);
     title(['Mode_' num2str(nm)], 'interpreter', 'tex', 'fontsize',
          22, 'fontweight', 'bold')
     grid off
     box on

     pause
     print(fig, [path 'Mode' num2str(nm) '.png'], '-dpng')
end

%% Make average structure

% Read in cartesian positions

disp(['Reading_positions_...'])

% Read in positions
h = read_dcdheader([path flname '_solu.dcd']);          % Ang
natoms = h.N;
nmols = 1;
nsteps = h.NSET;
startstep = 0;                                          % First step 0
step = 1;
laststep = nsteps-1;
nsteps_sel = ceil((laststep+1-startstep)/step);
pos = readdcd([path flname '_solu.dcd'], startstep, step, laststep,
     1:natoms);

posm = mean(pos, 1);
posmxyz = zeros(natoms, 4);
for aa = 1:natoms
    posmxyz(aa, 2:end) = posm(3*(aa-1)+1:3*(aa-1)+1+2);
end

% Write positions to file
fid = fopen([path flname '_mean.xyz'], 'wt');
formatSpec = '%s\t_%.6f\t_%.6f\t_%.6f\t_\n';
fprintf(fid, '%d\n', 38);
fprintf(fid, '\n');
for aa = 1:natoms
    fprintf(fid, formatSpec, syms{aa}, posmxyz(aa, 2), posmxyz(aa, 3)
         , posmxyz(aa, 4));
end
fclose(fid);

%% Write NMD file for Normal Mode Wizard

fid = fopen([path flname '_NMs.nmd'], 'wt');

% Title
fprintf(fid, [flname '\n']);

% Atom names
formatSpec = '%s';
fprintf(fid, ['names_']);
for aa = 1:natoms
    fprintf(fid, formatSpec, [syms{aa} '_']);
```

```
end
fprintf(fid, ['\n']);

% Residue names
formatSpec = '%s';
fprintf(fid, ['resnames ']);
for aa = 1:natoms
    fprintf(fid, formatSpec, [syms{aa} ' ']);
end
fprintf(fid, ['\n']);

% Coordinates
formatSpec = '%.6f %.6f %.6f ';
fprintf(fid, ['coordinates ']);
for aa = 1:natoms
    fprintf(fid, formatSpec, posmxyz(aa, 2), posmxyz(aa, 3), posmxyz(
        aa, 4));
end
fprintf(fid, ['\n']);

% NMs
for nn = 1:3*natoms
    fprintf(fid, ['mode ' num2str(nn) ' ']);
    fprintf(fid, '%.6f ', Lt(114-nn+1, :));
    fprintf(fid, ['\n']);
end

fclose(fid);
```

Listing A.4 Python script for extracting body-fixed frame velocities from ASE trajectory files.

```
#!/usr/bin/env python

import numpy as np
import os
import HPCPath as p

import rmsd

from sys import argv
from ase.io import read, write, Trajectory
"""
    Separetes translation, rotation and vibrations assuming
    there are no couplings.
    Rotates velocity vectors according to optimal superposition
    with respect to a reference structure.

    1) Translates frames such that COM coincides with COM
       reference structure

    2) Generate rotation matrix R to superimpose frames to
       reference  using Kabsch method
          (W. Kabsch, Acta Cryst. A 32, (1976))

    3) Rotates positions and velocities using R

                                      G. Levi 2017
"""

def get_angle(r1, r2, atom):
    v1 = r1[atom, :]
    v2 = r2[atom, :]
    v1n = v1 / np.linalg.norm(v1)
    v2n = v2 / np.linalg.norm(v2)
```

```python
    angle = np.arccos(np.vdot(v1n, v2n))

    return np.degrees(angle)


def get_vcom(v, masses):
    M = masses.sum()
    vcom = np.dot(masses.flatten(), v) / M

    return vcom

def get_etrans(vcom, masses):
    M = masses.sum()
    etrans = 0.5*M*np.linalg.norm(vcom)**2        # eV

    return etrans


def get_ang_velocity(atoms):
    """
        Sets the total angular momentum to zero
            by counteracting rigid rotations.
    """
    # Find the principal moments of inertia
        #  and principal axes basis vectors
    Ip, basis = atoms.get_moments_of_inertia(vectors=True)
    # Calculate the total angular momentum
        # and transform to principal basis
    Lp = np.dot(basis, atoms.get_angular_momentum())
    # Calculate the rotation velocity vector
        # in the principal basis, avoiding zero division
    # and transform it back to the cartesian coordinate system
        # Angular velocity in principal axis:
    omegap = np.select([Ip > 0], [Lp / Ip])
    omega = np.dot(np.linalg.inv(basis), omegap)
    # Compute rotational energy
    erot = 0.5*np.dot(omegap, Lp)
    # We subtract a rigid rotation
        # corresponding to this rotation vector
    positions = atoms.get_positions()
    vang = np.cross(omega, positions)

    return vang, erot


def get_rotation(P, Q):
    """
        Kabsch method to obtain a rotation matrix
        that minimizes the msd between an istantaneous
        structure and a static reference structure.
    """

    # Calculate covariance matrix
    cov = np.dot(np.transpose(P), Q)

    # SVD
    V, S, W = np.linalg.svd(cov)
    d = (np.linalg.det(V) * np.linalg.det(W)) < 0.0
    if d:
        S[-1] = -S[-1]
        V[:, -1] = -V[:, -1]

    # Generate rotation matrix
    R = np.dot(V, W)

    return R
```

```python
def multixyzwrite(ref, r, trajs, na, comr, syms, nameout, writeref=
    False):

    ref.set_positions(r)
    if writeref:
        # Write positions reference
        flref = open(pathout+'Reference.xyz','w')
        flref.write('%d\n' % na)
        flref.write('\n')
        for j in range(na):
            flref.write('%3s%14.6f%14.6f%14.6f\n' % (syms[j],r[j,0],r
                [j,1],r[j,2]))

        flref.close()
        # Write trajectory reference
        trajref = Trajectory(pathout+'Reference.traj', 'w', ref)
        trajref.write()
        trajref.close()

    # Move centroid to origin
    # Needed by the Kabsch method
    cref = r.mean(axis=0)
    ro = r - cref

    # Create trajectory object for dynamic frames
    trajr = Trajectory(pathout+nameout+'_solu.traj', 'w')

    ct = 0
    for tt, thistraj in enumerate(trajs):
        print 'Processing trajectory '+thistraj
        traj = Trajectory(thistraj)
        tl = len(traj)

        # Write header
        if tt == 0:
            flr = open(pathout+nameout+'_solu.xyz','w')
            flr.write('%d\n' % na)
            flr.write('Trajectory '+nameout+' Step 0\n')
            flrnoa = open(pathout+nameout+'_soluNoalign.xyz','w')
            flrnoa.write('%d\n' % na)
            flrnoa.write('Trajectory '+nameout+' Step 0\n')

            flv = open(pathout+nameout+'_vsolu.xyz','w')
            flv.write('%d\n' % na)
            flv.write('Trajectory '+nameout+' Step 0\n')
            flvnor = open(pathout+nameout+'_vsoluNorot.xyz','w')
            flvnor.write('%d\n' % na)
            flvnor.write('Trajectory '+nameout+' Step 0\n')

            fle = open(pathout+nameout+'_esolu.dat','w')
            fle.write('%3s%14s%14s%14s\n' % ('t', 'etrans', 'erot', '
                ekin'))

        for ii in range(tl):
            s = traj[ii][:na]
            s.set_constraint()
            fr = s.get_positions()
            fv = s.get_velocities()
            masses = s.get_masses()[:, np.newaxis]
            comf = s.get_center_of_mass()

            ### Positions
            # Translate origin to COM reference
            tvec = comf - comr
            fr -= tvec
```

```
# Write out positions before alignment
if ct is not 0:
        flrnoa.write('%d\n' % na)
        flrnoa.write('Step:_%d\n' %(ct))
for j in range(na):
        flrnoa.write('%3s%14.6f%14.6f%14.6f\n' % (
            syms[j],fr[j,0],fr[j,1],fr[j,2]))

# Move centroids to origin
# Needed by the Kabsch method
cr = fr.mean(axis=0)
fr -= cr

# Rotate to minimize rmd to reference
R = rmsd.kabsch(fr, ro)
frrot = np.dot(fr, R)

# Move centroids back
frrot += cr

## For an orthogonal (rigid) transformation det(R)=0
detR = np.linalg.det(R)

## Check RMSD
frrmsd = rmsd.rmsd(frrot, r)

print('Step:_' + str(ct) + ',_det(R):_' + str(detR) +
    ',_RMSD:_' + str(frrmsd))

# Write out positions
if ct is not 0:
        flr.write('%d\n' % na)
        flr.write('Step:_%d\n' %(ct))
for j in range(na):
        flr.write('%3s%14.6f%14.6f%14.6f\n' % (syms[j
            ],frrot[j,0],frrot[j,1],frrot[j,2]))


### Velocities
# Remove COM velocity
vcom = get_vcom(fv, masses)
fv -= vcom
etrans = get_etrans(vcom, masses)

# Set postions and velocities with respect to COM
                # before calculating angular velocity
s.set_positions(s.get_positions() - comf)
s.set_momenta(fv * masses)

# Remove angular velocity
vang, erot = get_ang_velocity(s)
fv -= vang
ekin = 0.5 * np.vdot(fv * masses, fv)

# Write out velocities before rotation
if ct is not 0:
        flvnor.write('%d\n' % na)
        flvnor.write('Step:_%d\n' %(ct))
for j in range(na):
        flvnor.write('%3s%14.6f%14.6f%14.6f\n' % (
            syms[j],fv[j,0],fv[j,1],fv[j,2]))

# Rotate to reference
fv = np.dot(fv, R)

# Write out velocities
```

```
                              if ct is not 0:
                                   flv.write('%d\n' % na)
                                   flv.write('Step: %d\n' %(ct))
                              for j in range(na):
                                   flv.write('%3s%14.6f%14.6f%14.6f\n' % (syms[j
                                       ],fv[j,0],fv[j,1],fv[j,2]))

                              # Write to trajectory
                              s.set_positions(frrot)
                              s.set_velocities(fv)
                              trajr.write(s)

                              # Write out energies
                              fle.write('%3d%14.6f%14.6f%14.6f\n' % (ct*2, etrans,
                                  erot, ekin))

                              ct = ct + 1

     flr.close()
     flrnoa.close()
     flv.close()
     flvnor.close()
     fle.close()
     trajr.close()

##############################################################

### Define variables
na = 38                       # Number of atoms in solute
pathout = p.HPCPath().path+'NMs/'

### Get inputs
del argv[0]

# Read in reference structure
# Reference is first frame of trajectory in first argument
trajref = Trajectory(argv[0])
ref = trajref[0][:na]
ref.set_constraint()
r = ref.get_positions()
syms = ref.get_chemical_symbols()
comr = ref.get_center_of_mass()

# Read in trajectories to process
trajs = []
for filename in argv[1:]:
     if '.traj' in filename:
        trajs.append(filename)

# Read name for output file if given
if ('.traj' not in argv[-1]):
   nameout = argv[-1]
else:
   nameout = trajs[0][0:-5]

multixyzwrite(ref, r, trajs, na, comr, syms, nameout, writeref=False)
```

Appendix B
Further Details on the Vibrational Analysis in Solution

Figure B.1 shows the main generalized normal modes involved in the vibrational relaxation of PtPOP in water, as obtained from the vibrational analysis of the S_1 solution-phase trajectories of the second set of ΔSCF-QM/MM BOMD simulations performed in the present work. The pinching mode has almost exclusive character of Pt-Pt stretching and a period of $\sim$236 fs. Thus, it is very similar to the pinching mode extracted from the vacuum ΔSCF-QM BOMD simulations (compare with Fig. 13.5). The mode indicated as mode a has prevalent character of asymmetric Pt-P stretching, with an FT of mode velocities peaking at $\sim$120 fs. Mode b and mode c have large overlaps with, respectively, the twist 1 and breathing modes obtained from the generalized normal mode analysis of the vacuum trajectories. Their characteristic periods are also very close to those of the vacuum twist 1 and breathing modes (compare with Fig. 13.5). Notably, mode b and mode c present significant contributions from Pt-P stretchings in addition to characters of twisting and breathing.

© Springer Nature Switzerland AG 2019
G. Levi, *Photoinduced Molecular Dynamics in Solution*, Springer Theses,
https://doi.org/10.1007/978-3-030-28611-8

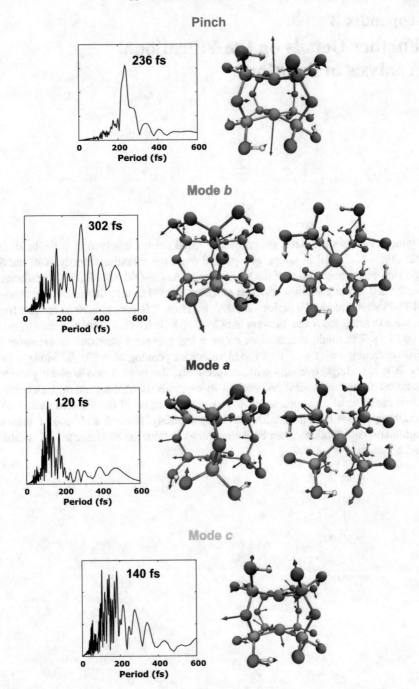

Fig. B.1 The pinching mode and the modes that were found to be more coupled to it from the generalized normal mode analysis of the ΔSCF-QM/MM BOMD simulations of PtPOP in water. The modes are represented through generalized normal mode displacement vectors. For each of them the FT of the autocorrelation function of mode velocities is also shown

Printed in the United States
By Bookmasters